Emilienne Tudor Ngo Sock

Clerodendrum umbellatum may treat Schistosoma mansoni schistosomiasis

Emilienne Tudor Ngo Sock

Clerodendrum umbellatum may treat Schistosoma mansoni schistosomiasis

ScienciaScripts

Imprint

Any brand names and product names mentioned in this book are subject to trademark, brand or patent protection and are trademarks or registered trademarks of their respective holders. The use of brand names, product names, common names, trade names, product descriptions etc. even without a particular marking in this work is in no way to be construed to mean that such names may be regarded as unrestricted in respect of trademark and brand protection legislation and could thus be used by anyone.

Cover image: www.ingimage.com

This book is a translation from the original published under ISBN 978-3-639-60828-1.

Publisher:
Sciencia Scripts
is a trademark of
Dodo Books Indian Ocean Ltd. and OmniScriptum S.R.L publishing group

120 High Road, East Finchley, London, N2 9ED, United Kingdom
Str. Armeneasca 28/1, office 1, Chisinau MD-2012, Republic of Moldova, Europe
Managing Directors: Ieva Konstantinova, Victoria Ursu
info@omniscriptum.com

Printed at: see last page
ISBN: 978-620-9-76034-1

TABLE OF CONTENTS

DEDICATION

I dedicate this work to my entire family

ACKNOWLEDGMENTS

I would like to express my deepest gratitude to all those who, in one way or another, contributed to the completion of this work. In particular, I would like to thank:

- God, who has continually granted me the health, strength, and courage necessary to complete this work;

- to the FIS (IFS Grant F/3622-1) for funding this project, without which this work would not have been possible;

- Dr. Hermine BOUKENG JATSA, née MEGAPTCHE, for supervising this work;

- Prof. Pierre KAMTCHOUING for supervising this work;

- to Prof. Louis-Albert TCHUEM TCHUENTE, for welcoming me to the "Center for Schistosomiasis & Parasitology," of which he is the director. His support enabled me to work under favorable conditions, for which I am particularly grateful;

- to the student researchers at the Laboratory of Bioorganic and Medicinal Chemistry in the Faculty of Sciences at the University of Yaoundé I for their contribution to conducting the phytochemical tests on the extract used in this study. I would particularly like to thank Mr. Jules NONO and Mr. Yves NGANSO;

- to Mr. Harquin FOYET of the Animal Physiology Laboratory at the Faculty of Sciences, University of Yaoundé I, who provided the reagents necessary for conducting the phytochemical tests for this thesis;

- to Drs. Désiré DZEUFIET, Léonard TEDONG, and Paulin NANA, who contributed to the biochemical and histopathological analyses for this thesis;

- to all my classmates for their moral support, camaraderie, and advice;

- to Mr. and Mrs. MBOUE for kindly welcoming me into their home and treating me like their own child throughout the entire period of this research;

- to all my family members and friends who contributed to the completion of this work.

- There are many who, through a simple word of encouragement or a positive thought, have carried me in their hearts and contributed to my success through their prayers. The omission of their names should not be considered an oversight, much less a sign of contempt or dis . May they feel particularly acknowledged by this paragraph and accept my sincere thanks.

SUMMARY

Clerodendrum umbellatum is a plant belonging to the Lamiaceae family. Its leaves have been used for decades in traditional medicine to treat intestinal helminthiasis, epilepsy, asthma, headaches, irregular menstruation, and dermatitis. The objective of this study was to test the efficacy of the aqueous leaf extract of *Clerodendrum umbellatum* against *Schistosoma mansoni* schistosomiasis in male BALB C mice. Infested male mice were treated orally with the aqueous leaf extract of *Clerodendrum umbellatum* at doses of 40, 80, 160, and 320 mg/kg. Treatment with this extract lasted 14 days (from the^{60th}day post-infestation to the^{74th}day). After 7 days of observation, the mice were sacrificed and several parameters were evaluated. The resulting findings are as follows: From a parasitological perspective, the following were noted on the^{60th}day post-infestation: the presence of eggs in the feces and liver, the presence of schistosomes in the blood vessels, a decrease in body weight, hepatomegaly and splenomegaly, as well as granulomas on histological sections of the liver. Treatment with the aqueous extract of *Clerodendrum umbellatum* leaves resulted, in the treated animals as a whole, in a reduction in the average egg load in feces by 54.48% and in the liver by 22.74%, as well as a significant decrease in the parasite load by 88.45%. Hepatomegaly and splenomegaly were reduced after administration of the extract. Although granulomas were still visible on histological sections in mice treated with certain doses, their sizes were nevertheless reduced. In terms of reproduction, *Schistosoma mansoni* infection is responsible, by the 81thday post-infestation, a decrease in the weight of the testes, epididymis, and seminal vesicles, as well as in epididymal sperm concentration and epididymal alpha-glucosidase activity. However, there was an increase in vesicular fructose levels. Treatment of the mice with an aqueous extract of *Clerodendrum umbellatum* leaves led to an improvement in these parameters, bringing them closer to those of healthy mice. Phytochemical analysis of the *Clerodendrum umbellatum* extract revealed the presence of alkaloids, flavonoids, saponins, saponosides, tannins, and triterpenes. The repair of abnormalities caused by the infection could be explained by the reduction in egg load and parasite load, a reduction resulting from the presence in the extract of bioactive compounds with antiparasitic, antioxidant, and anti-inflammatory activity. In conclusion , the aqueous extract of *Clerodendrum umbellatum* leaves has an anti-schistosomiasis effect, which results in an improvement in the parameters studied.

Introduction

Schistosomiasis, also known as bilharzia or schistosomiasis, is a parasitic disease caused by the presence of adult worms (schistosomes) in the blood vessels. They are caused by flatworms of the class Trematoda, genus *Schistosoma*, whose infective forms (furcocercariae) actively penetrate the body through the skin (Ollivier et al. 1998). The parasitic life cycle involves a mammal and a specific freshwater mollusk, which lives in its habitat where it is infested by miracidia and releases infective cercariae.

According to statistics, schistosomiasis is the second leading cause of parasitic morbidity in tropical and subtropical countries after malaria. This disease is found in at least 74 developing countries, where approximately 600 million people are at risk, over 200 million are infected, and 20 million suffer from the severe consequences of the disease (WHO, 1997). In Cameroon, 1.7 million people are affected (Brooker et al., 2000). The northern provinces of the country are the most affected by this disease (Ratard et al., 1990).

Despite advances in medicine, schistosomiasis remains a public health problem. Control of the disease continues to rely on treatment with the drug of choice, praziquantel, for all forms of schistosomiasis. However, previous studies report that praziquantel causes hemorrhage in the host's lung tissue (Flisser and Mclaren, 1989) as well as abdominal pain and diarrhea (Katabareine et al., 2003). Studies also indicate a risk of certain schistosome strains developing resistance to this drug, primarily due to the problem of reinfection following treatment (Silva et al., 2003), hence the need to search for alternative drugs for the treatment of schistosomiasis. This finding led to the establishment of a project titled "Search for natural compounds with anti-schistosomal activity in Cameroonian medicinal plants." To this end, certain species of Cameroonian medicinal plants used to treat intestinal helminthiasis were selected. *Clerodendrum umbellatum*, which is part of this selection and on which we have focused our attention, is a plant of the Labiatae family. The leaves of this plant are used in traditional medicine to treat conditions such as intestinal helminthiasis, epilepsy, asthma, headaches, irregular menstruation, and dermatitis (Adjanohoun et al., 1996). Our main objective in this study was to test the treatment of *Schistosoma mansoni* schistosomiasis using an aqueous extract of *Clerodendrum umbellatum* leaves and to investigate the impact of the disease and treatment on several parameters of male reproductive function in BALB C mice. To achieve this objective, we:

- studied the impact of *Schistosoma mansoni* schistosomiasis on the liver, spleen, and male reproductive function;

- examined the effect of treating *Schistosoma mansoni* schistosomiasis with an aqueous extract of *Clerodendrum umbellatum* leaves;

- and evaluated the effect of treatment for *Schistosoma mansoni* schistosomiasis on the liver, spleen, and certain parameters of male reproductive function.

LITERATURE REVIEW

I- DESCRIPTION OF *CLERODENDRUM UMBELLATUM* POIR

Also known as *Clerodendrum scandens* P. Beauv.; Nganwi in Bafut; Biyem or Dion in Bassa; Eloi-dibi or Elok dibi in Ewondo; Dor nkol in Kaka, *Clerodendrum umbellatum* is a plant of the Verbenaceae family (Adjanohoun et al., 1996), or the Labiatae (Cheek et al., 2004).

I-1- Systematics (Cheek et al., 2004)

Kingdom:	Plantae
Subkingdom:	Vascular plants
Division:	Spermatophytes
Subdivision:	Angiosperms
Class:	Dicotyledons
Order:	Lamiaceae
Family:	Labiatae
Genus:	*Clerodendrum*
Species:	*Clerodendrum umbellatum*

I-2- Description

I-2-1- General Characteristics of the Verbenaceae

This family includes shrubs or small trees. The leaves are opposite, simple, or compound. Stipules are absent; the hermaphroditic flowers are zygomorphic; the calyx is lobed or toothed. The gamopetalous corolla is tubular and lobed, with the lobes imbricated. There are 4 stamens on the corolla, rarely 2 or 5. The anthers, which are most often divergent, bear the gametes. The superior ovary contains 2 to 8 reproductive cells. The terminal style is simple; the fruit contains seeds with a straight, constricted embryo (Hutchinson and Dalziel, 1963).

I-2-2- *Clerodendrum umbellatum*

It is a shrub about 7 m tall with long, slender branches. The opposite, leaves are elliptical, about 6 to 20 cm long and 3.5 to 11 cm wide. The terminal inflorescence contains white flowers, sometimes pink at the base; the serrated calyx is green at the bottom and white at the top; the fruit splits into 4 parts when it reaches maturity. This plant is found on the African continent, specifically in Senegal, Congo, Sudan, and Ethiopia (Adjanohoun et al., 1996; Cheek et al., 2004).

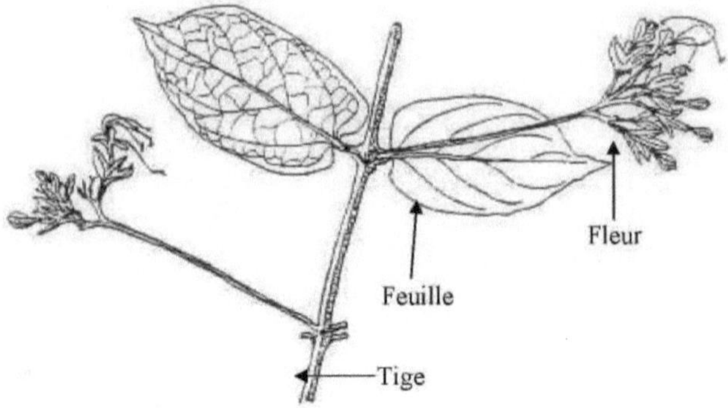

Figure 1: Morphological characteristics of *Clerodendrum umbellatum* (Adjanohoun et al., 1996)

I-3- Ethnopharmacological study

The leaves of *Clerodendrum umbellatum* are used in traditional medicine to treat conditions such as epilepsy, asthma, headaches, irregular menstruation, dermatitis, and intestinal helminthiasis (Adjanohoun et al., 1996).

The treatment for intestinal helminthiasis involves grinding a sufficient quantity of fresh *Clerodendrum umbellatum* leaves in one liter of water, drinking half of the filtrate, and purging with the other half after adding five *Capsicum frutescens* fruits (Adjanohoun et al., 1996).

I-4- Chemical Study

No chemical studies have yet been conducted on the species *Clerodendrum umbellatum*.

However, chemical compounds have been isolated from several species of the genus *Clerodendrum*. This is the case for the flavonoids isolated from the ethanolic extract of *Clerodendrum petasites*, which appear to have a bronchodilator effect (Hazekamp et al., 2001); , isoacteoside, and Jionoside D, which has antioxidant properties, isolated from

Clerodendrum trichotomum (Chae et al., 2004; Chae et al., 2005). This is also the case for the cyanogenic glycosides extracted from *Clerodendrum grayi* (Miller et al., 2006) and the phenylethanoid glycoside extracted from *Clerodendrum inerme* (Nan et al., 2005).

II- GENERAL INFORMATION ON SCHISTOSOMIASIS CAUSED BY *SCHISTOSOMA MANSONI*

II-1- Definition, Classification, and Distribution

- Definition

Bilharzia, also known as schistosomiasis or schistosomosis, are parasitic diseases caused by the presence of adult worms (schistosomes) in the blood vessels (Ollivier et al. 1998).

- Classification (Sellin and Boudin, 1981; Southgate et al., 1998).

Phylum:	Plathelminthes
Class:	Trematodes
Subclass:	Digenia
Superorder:	Anepitheliocystidia Larue
Order:	Strigeatoidea Larue
Suborder:	Strigeata Larue
Superfamily:	Schistosomatidea
Family:	Schistosomatidae (Laos, 1899)
Subfamily:	Schistomatinae (Stiles and Hassal, 1926)
Genus:	*Schistosoma* (Weinland, 1858)
Species:	*Schistosoma mansoni* (Sambon, 1907)

- Distribution

There are six species of schistosomes that are pathogenic to humans (**Table 1**). Three of these species are found on the African continent.

- *Schistosoma haematobium* Bilharz 1852: causative agent of urogenital schistosomiasis. This parasite is found throughout Africa, in Madagascar, and in the Near and Middle East

(Golvan, 1983). In Cameroon, it is found in the Far North, North, Littoral (Loum and Penja), Southwest (Kumba and Barombi Kotto), and West (Kékem) provinces (Ratard et al., 1990). Transmission is mediated by intermediate host snails of the genus *Bulinus* (Golvan, 1983).

- *Schistosoma intercalatum* Fisher 1934 develops on snails of the *Bulinus africanus* group, and its geographic distribution appears to be limited to the Zaire River basin. The new taxon *S. guineensis* should be reserved for the agent of rectal schistosomiasis that develops in its larval form exclusively in snails of the *Bulinus forskalii* group: the species is believed to be distributed around the Gulf of Guinea (Cameroon, Gabon, Equatorial Guinea, São Tomé) and is reportedly absent from the Zaire Basin (Pagès et al., 2003).

- *Schistosoma mansoni* Sambon 1907, responsible for intestinal and sometimes hepatosplenic schistosomiasis. This species is widely distributed across the African continent. Its range also includes Madagascar, intertropical America, and the West Indies (Malek, 1980). In Cameroon, this schistosomiasis is endemic mainly in the North and Adamaoua provinces, with extensions into Mayo Tsanaga in the Far North and the regions of Makénéné, Gah Bape, Bafia, Bokito, Ndikinimeki, and Yaoundé in the Central Province (Ratard et al., 1990). The intermediate host mollusks are Planorbinae pulmonates of the genus *Biomphalaria* (Golvan, 1983; Bourée, 1989).

Table: Main characteristics of schistosomes parasitizing humans (Tchuem Tchuenté, 1993)

Egg morphology **Species**	**Vector mollusks**	**Geographic distribution**
-Eggs with a terminal spur:	Aquatic pulmonates of the genus *Bulinus*	
5. Haematobium	***B. globosus, B. truncatus ...***	Africa, Madagascar, Near and Middle East, India
S. intercalatum	***B. forskalii***	Central Africa
-Eggs with lateral spurs:	Aquatic pulmonates: genus *Biomphalaria*	
S. mansoni	***B. pfeifferi, B. glabrata....***	Africa, Madagascar, Intertropical America, West Indies
-Eggs with a rudimentary spur:	Prosobranchs Pomatiopsidae	
S. japonicum	***Oncomelania hupensis***	Mainland China and the Southeast Asian archipelagos
S. mekongi	***Tricula aperti***	Asian
S. malayensis	***Robertsiella kaporensis***	Laos, Thailand, Cambodia,

		Malaysia

II-2- Biology of the parasite

II-2-1- Life cycle (Figure 2)

The stages of this cycle vary little from one species to another; only the intermediate host mollusks differ.

In the definitive host, the worms form permanent pairs that settle in the mesenteric veins and reproduce sexually. The lifespan of schistosomes in humans is estimated to be 2 to 20 years, or even up to 35 years (Combes et al., 1987). Each mated female lays approximately 300 eggs per day (this can exceed 3,000 eggs in *S. japonicum*). However, only a small proportion (20–30%) of the eggs laid in the venules manage to penetrate the intestinal or bladder wall, depending on the species, and are eliminated with the feces (1). An aquatic environment is essential for the continuation of the parasite's life cycle.

In the aquatic environment, the egg (2) releases a swimming ciliated larva, the miracidium (3), which carries out the first phase of the parasite's dispersal. Upon encountering the specific mollusk, the miracidium penetrates it and becomes the primary sporocyst (4), which, through asexual reproduction, produces numerous daughter sporocysts (5). These colonize the mollusk's digestive gland and produce, in large numbers, cercariae (6) that escape from the host. The intense multiplication of the parasite within the mollusk (sporocystogenesis and cercariogenesis) is substantial; from a single miracidium, several thousand cercariae are produced.

The cercaria, a furcocercaria-type larva, represents the second stage of the parasite's dispersal in the aquatic environment. It has only a few hours to find its target. When the cercaria encounters a suitable vertebrate host, it penetrates the host's skin (7). This penetration is complete in about 10 minutes. As they penetrate the dermis, the cercariae shed their tails and transform into schistosomulae. These enter the bloodstream, which carries them to the hepatic portal vein via the heart and lungs, where they feed on red blood cells, grow, differentiate into males and females, and mate. The migratory phase lasts a total of 7 to 21 days. Males carrying femal s form pairs (8) that actively swim upstream in the portal vein and enter the mesenteric veins, where the eggs are laid. These eggs are the primary cause of disease in the vertebrate host (Tchuem Tchuenté, 1993).

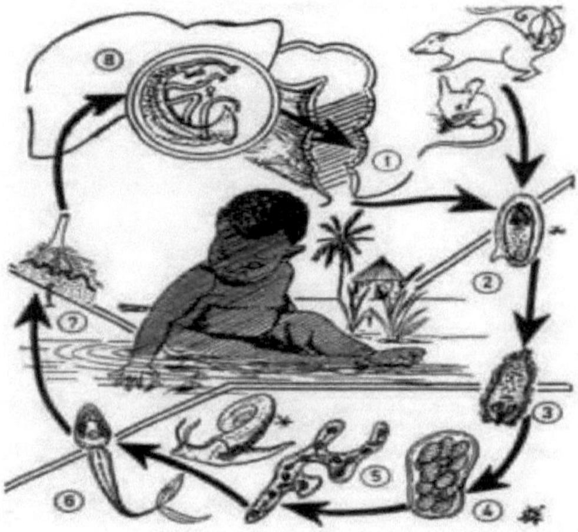

Figure 2: Diagram of the life cycle of human schistosomes (here *Schistosoma mansoni*) (Golvan, 1983).

1 : excretion of eggs in feces, **2**: egg, **3**: miracidium, **4**: sporocyst, **5**: daughter sporocyst, **6**: cercaria, **7**: penetration of cercariae into the human host, **8**: adults in the liver.

II-2-2- Pathogens (Figure 3)

- Schistosomes have a similar morphology across species. The larger male (13 x 0.7 mm) houses a thread-like female (17 x 0.1 mm) in a longitudinal groove (the gynaecophoric canal) (Bourée, 1989). They are equipped with two suckers: one oral and the other ventral. The oral sucker facilitates feeding by absorbing blood, which is digested in the digestive tract. This tract, which lacks a pharynx, begins with a short esophagus that divides into two intestinal loops merging posteriorly into an axial cul-de-sac. An excretory system and genital organs are also present. In males, the number of testicles varies by species. In females, the number of mature eggs present in the uterus ranges from one to over a hundred, but this number bears no relation to the number of eggs that can be laid in 24 hours (Golvan, 1983).

- The egg has a characteristic spur and contains a miracidium. Its shape and size vary by species: In *S. mansoni*, it is oval (110 x 60 µm) and has a lateral spur; in *S. intercalatum*, it is also oval (200 x 70 µm) but has a terminal spur; in *S. japonicum*, it is nearly round (70 x 50 µm) and has a very short spur; that of *S. mekongi* is similar to the previous one, with a few differences in dimensions (65 x 55 µm), and in *S. haematobium*, its shape is oval (150 x 60 µm) with a terminal spur (Bourée, 1989).

- The miracidium is a small ciliated organism measuring 200 to 250 µm. It has a rudimentary digestive tract and a few excretory cells with a vibratile flagellum. It moves fairly quickly in a straight line and can survive for about 24 hours under favorable environmental conditions (Golvan, 1983).

- The cercaria, approximately 500 μm long, has an anterior portion (the head) that is oval and elongated, with two suckers and a rudimentary bifurcated digestive tract. Small anterior spines likely play a role in perforating the definitive host's integument, thereby aiding the action of the lytic secretions from the cephalic glands. This anterior portion is connected to a bifurcated tail by a fragile attachment that allows it to detach during penetration. This distinctive shape has earned it the name furcocercaria (Golvan, 1983).

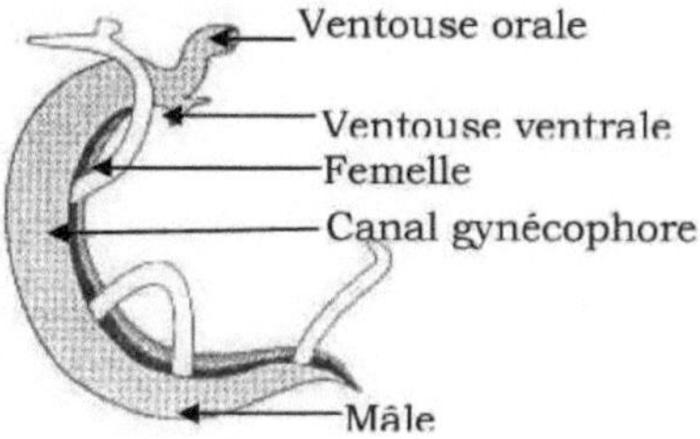

a- Pair of schistosomes (here, *S. mansoni*) (Théron et al., 1992)

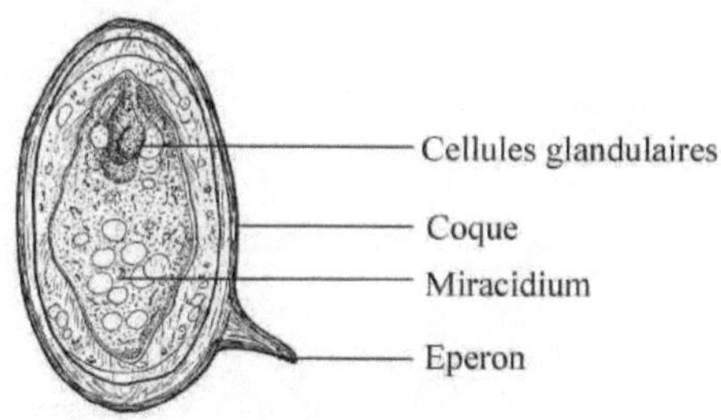

b- Egg of *S. mansoni* (Golvan, 1983)

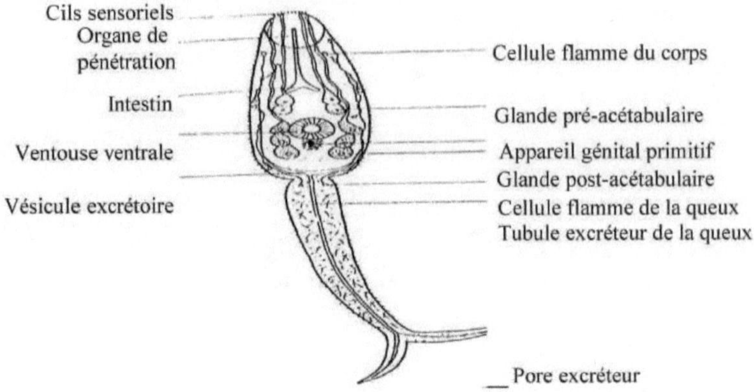

c- Cercaria of *S. mansoni* (Malek, 1980)

<u>Figure 3</u>: Illustration of a pair of schistosomes, the egg, and the cercaria of *Schistosoma mansoni*

II-3- The disease: *Schistosoma mansoni* schistosomiasis

II-3-1- Pathology of schistosomiasis

Two modes of action underlie the pathological manifestations in schistosomiasis:

- the release into the host's body of foreign substances (free radicals and toxic substances) resulting from the parasite's metabolism. These foreign substances are responsible for all the clinical manifestations of the early stages of schistosomiasis (Houin, 1981);

- during their trans-tissue migration, the eggs also release such substances, which are responsible for the formation of schistosomal granulomas (bilharziomas), elementary lesions whose accumulation gives rise to all the characteristic features of the chronic phase (Grossetete et al., 1989).

II-3-2- Clinical Manifestations

Chronologically, the clinical manifestations of schistosomiasis can be grouped into three phases of varying importance: the immediate phase of skin reactions to the penetration of furcocercariae; the invasion phase occurring after an incubation period; the chronic phase marked by the symptoms of the specific schistosomiasis and progressing toward complications (Evengard, 1989).

II-3-2-1- Skin Reactions to Infestation

These reactions reflect the penetration of cercariae through the skin and their migration into the integument. The most obvious sign is pruritus, which begins 15 to 30 minutes after the contaminated bath. Subsequently, diffuse erythema or urticarial papules may appear, sometimes progressing in successive outbreaks over two to three days (Golvan, 1983). Reinfestations with human schistosomes are asymptomatic, at least if the initial schistosomiasis has not been cured.

II-3-2-2- Invasion phase

During this phase, digestive symptoms appear to be the most prominent. The patient complains of abdominal pain, headaches, diarrhea, nausea, and vomiting. Clinically, hepatosplenomegaly and hypereosinophilia are observed (Nash et al., 1982). Fever may reach 39 to 40°C, accompanied by chills, headaches, muscle or joint pain, and profuse sweating. The appearance of hives and edema about ten days after this toxic-infectious onset directs the investigation toward helminthiasis. In a significant number of cases, pulmonary congestion is also observed, as the presence of schistosomula in the lungs is irritating and induces a dry cough (Houin, 1981; Golvan, 1983).

II-3-2-3- Chronic Phase

At this stage, the manifestations are no longer unambiguous, as they depend on the egg-laying sites of each species. For *S. mansoni*, for example, egg-laying occurs in the branches of the portal system, which explains the location of the lesions: intestine, liver, and spleen.

This phase is dominated by intestinal signs. Most patients from endemic areas complain of colic: postprandial bloating exacerbated by high-fiber foods, and bowel dysfunction characterized by brief episodes of diarrhea against a background of chronic constipation. Among immigrants in endemic areas, clinical presentations resembling the subacute forms of intestinal amebiasis may be observed, with pain along the colonic region, frequent stools (about ten per day), loose, and sometimes mucous. The condition progresses toward sclerosis of the colon wall, leading to hypersecretion of mucus and the occurrence of painfully chronic episodes, especially during periods of fatigue or following dietary indiscretions (Golvan, 1983). The presence of eggs in tissues and organs ( ) leads to the formation of schistosomal granulomas. Each granuloma consists of an egg surrounded by eosinophils, histiocytes, epithelial cells, and fibroblasts (Gentillini et al., 1982).

II-3-2-4- Complications (Golvan, 1983)

- Hepatosplenic manifestations: Initially, the liver is congested, smooth, firm, and sometimes tender to the touch. The spleen is also enlarged. Eggs are already present in the liver parenchyma. During the established phase, hepatic fibrosis begins. The patient's general condition deteriorates progressively as the disease advances. Hepatic fibrosis is initially hypertrophic and then becomes atrophic. The patient loses weight, develops a fever, the complexion turns sallow, and the mucous membranes lose their color, indicating anemia. Portal hypertension is evident and is due to a presinusoidal blockage, whereas suprahepatic pressure is normal. The liver becomes hard and is dotted with whitish spots, which are Schistosoma granulomas. Splenomegaly, meanwhile, is constant; sometimes a form is present in which splenic involvement is the primary feature. Indeed, this manifests as a progressive increase in the size of the spleen, which becomes hard and reaches monstrous proportions. Signs of hypersplenism are evident, and it is these, along with the rupture of esophageal varices, that account for the significant risk of hemorrhage threatening the future of these patients. In the longer term, these hepatic or splenic manifestations will lead to liver failure. The prognosis for these complications is therefore grim, and this is also true when they occur with other species of schistosomes.

- Rectocolonic polyposis: Small growths may form on the colonic wall around clusters of eggs. The rectum is generally the site of choice for these "polyps," which cause a sensation of fullness in the ampulla, triggering the defecation reflex.

- Cardiac and pulmonary forms: The invasion phase may be marked by acute pulmonary manifestations such as febrile edema, or bronchopneumonia with purulent or bloody sputum and severe impairment of general health. Complications of the hepatosplenic form include pulmonary arteritis and right ventricular hypertrophy.

- Nervous system manifestations: These are much rarer but severe. Embolic eggs can become lodged at any point along the central nervous system and cause a wide range of motor, sensory, or sensory-motor symptoms. Most often, the presentation is that of transverse myelitis, which is difficult to attribute to schistosomiasis until it is discovered during a comprehensive examination.

II-3-2-5- *Schistosoma mansoni* schistosomiasis and reproductive function

With regard to the male genital tract, observations from autopsies have revealed the presence of *S. mansoni* and *S. haematobium* eggs in the prostate, seminal vesicles, and vas deferens (Gelfand et al., 1970; Patil and Elem, 1988). This trapping of eggs in the seminal vesicles and prostate leads to calcification of the tissues of these organs, a process that results in prostate adenocarcinoma in a limited number of patients (Ma and Srigley, 1995; Vilana et al., 1997). A study analyzing the semen of two schistosomiasis patients revealed azoospermia; testicular biopsy showed inhibition of spermatogenesis and the presence of schistosome eggs in the interstitial tissue (Girgis and Wassef, 1980). Clinical data report testicular infarction caused by *S. mansoni* (Steinberger et al., 1975).

Hepatic complications lead to increased levels of serum estrone and estriol, while serum estradiol levels are normal or decrease slightly (Rizk et al., 1980). As for testosterone, Cavalière et al. (1986) and El-Ridi et al. (1987) show that its levels are reduced; studies by Abdalla et al. (1994) report an increase, whereas Skelly et al. (1994) find unchanged serum levels in *S. mansoni* schistosomiasis patients. Furthermore, a decrease in serum FSH levels and phospholipid levels, as well as an increase in triglycerides and cholesterol in the testis, have been reported (Saleh and Hamdy, 1979).

In women with *Schistosoma mansoni* schistosomiasis, a reduction in ovarian weight and atrophy of the corpus luteum followed by a halt in its development have been observed (Tiboldi, 1979). Data in the literature are contradictory regarding the hormonal profile of females infected with *S. mansoni*. Tiboldi et al. (1979) note a reduction in serum progesterone levels, whereas Abdalla et al. (1994) instead observe an increase in progesterone and 17β-estradiol levels. Regarding the fertility of female mice infected with *S. mansoni*, Amano et al. (1990) report a reduction in the number of viable litters, resulting from the death of parents during gestation, abortion, and infanticide. A reduction in body weight was also observed in the offspring of infected mice at two weeks of age.

II-3-3- Diagnosis

This involves parasitological examination of the patient's stool, rectal biopsy, and immunological tests.

S. mansoni eggs are found in greater quantities in mucoid-bloody stools, and the best chance of detecting them under a microscope is by scraping the surface of the stool coated with mucus (Ripert, 1998).

Analysis can be qualitative (detection of eggs in stool) or quantitative (egg counting). The Kato-Katz method (Katz, 1970), recommended for epidemiological surveys, yields satisfactory results. The concentration method using the MIF (merthiolate-iodine-formol) technique (Golvan, 1983) and the concentration method (centrifugation), which involves diluting the stool, sieving and filtering the resulting solution through several sieves with different mesh sizes, then centrifuging the filtrate and collecting the pellet for microscopic examination, are also used (Golvan, 1983).

Rectal biopsy is a very high-quality examination. The eggs are clear, and it is often possible to see the miracidium moving within its shell. The sample is obtained via a simple anoscopy, from the anterior wall of the ampulla, or preferably from the rim of a Houston valve. It is necessary to obtain not only mucosa

but also the submucosa, since that is where the eggs are laid. The biopsy specimen can be thinned and preserved in chloral gum. Rectoscopy appears to be indicated only for assessing polypoid lesions, for biopsying these tumors, and for detecting possible malignancy.

For immunological assays, immunofluorescence is used on liver sections of *S. mansoni*. The assay is also quantitative and allows for monitoring the progression of schistosomiasis (Golvan, 1983).

II-3-4- Control

There are numerous methods for controlling schistosomes, and they vary depending on the stage of the parasite's life cycle being targeted.

- In the natural environment, ecological methods include the sanitation of stagnant or flowing water bodies that serve as mollusk habitats; regular fluctuations in water levels through periodic draining of irrigation canals hinder the reproduction of mollusk- ; and chemical control of intermediate hosts through the application of molluscicides. This latter control method is not widely favored, as these products also destroy non-target species such as fish and crustaceans, which can disrupt the ecological balance. A control method preferable to the previous one is biological control, which involves introducing into the water containing the mollusks, for example, insects of the Sciomyzidae family that prey on mollusks

(Woolhouse and Chandiwana, 1990). Predators such as ducks, fish (such as certain cichlids), insects (aquatic larvae of Sciomyzidae flies, naucores, diving beetles, bellostomes), and leeches have been tested and have proven effective (Golvan, 1983).

- In the definitive host, treatment of patients with appropriate medications (chemotherapy) such as praziquantel and oxamniquine (Gentillini et al., 1982; Evangard, 1989).

Some authors recommend the use of medicinal plants to combat schistosomiasis. Hamed and Hetta (2005) recommend the use of *Citrus reticulata* and Mirazid because they significantly reduce the number of eggs and schistosomes. However, studies by other authors on Mirazid have instead cast doubt on its efficacy against schistosomiasis. These include the studies by Botros et al. (2004 and 2005), as well as those by Rashida et al. (2005).

Preventive measures are also effective; at the individual level, this involves protecting submerged body parts when working in water; at the community level, health education for populations living in endemic areas must be effectively implemented (Golvan, 1983). It is preferable to raise awareness and educate populations living in endemic areas about fecal hygiene and the dangers of bathing in infested backwaters.

Previous studies have shown that schistosome eggs can be found in the male genital tract and consequently affect reproductive organs such as the testicles and accessory glands. In the liver and spleen, the presence of these eggs leads to hepatomegaly and splenomegaly in the case of *Schistosoma mansoni*.

III- GENERAL INFORMATION ON THE SEXUAL GLANDS

III-1- The Testis

III-1-1- Structure of the Testis (Figure 3)

The mammalian testis is an organ covered by the tunica albuginea, a thick fibrous membrane. It comprises two distinct compartments (Marieb, 1999):

- an avascular exocrine compartment accounting for 90% of the testis' total mass. It contains germ cells in various stages of development, Sertoli cells, and peritubular or myoid cells belonging to the *tunica propria*, a complex structure separating the seminiferous tubules from the interstitial tissue. This compartment is directly connected to the epididymis via the *rete testis* and forms a functional unit necessary for the maturation and storage of spermatozoa.

- an endocrine compartment consisting of a highly vascularized interstitial space containing Leydig cells, blood and lymphatic vessels, macrophages, and fibroblasts.

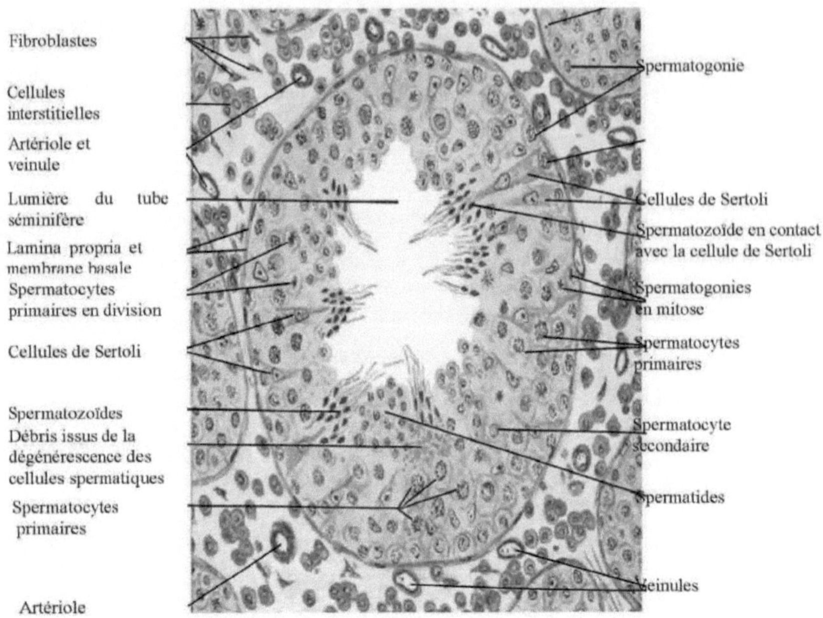

Figure 3: Histological section of a seminiferous tubule (panoramic view) x 300 (Di Fiore, 1965)

III-1-2- Functions of the Testis

The testis has a dual function due to its structural organization: exocrine (spermatogenesis) and endocrine (steroidogenesis).

III-1-2-1- Exocrine function: spermatogenesis

Spermatogenesis is the series of events that take place in the convoluted seminiferous tubules and lead to the production of male gametes, the spermatozoa. This process begins in boys at puberty and continues throughout life (Grizard and Jimenez, 1997).

- **Germ cells**

During spermatogenesis, three main cell types are involved: spermatogonia, spermatocytes, and spermatids.

Spermatogonia are cells located at the periphery of the seminiferous tubules, between the Sertoli cells. During the first phase of spermatogenesis, these spermatogonia multiply through normal somatic mitosis. These divisions allow for the renewal of spermatogonia as well as the differentiation of these cells into spermatocytes I (diploid cells). The second phase of spermatogenesis, which corresponds to meiosis, is divided into two stages: a reductive mitosis that leads to a halving of the genetic material, during which spermatocytes I divide to produce spermatocytes II (haploid cells); these spermatocytes II then undergo an equational mitosis to produce spermatids (haploid cells). In the final phase, which constitutes spermiogenesis, the spermatids undergo a series of profound morphological transformations to ultimately give rise to spermatozoa. The total duration of spermatogenesis is 74 days in humans and 35 days in mice. At the end of spermatogenesis, the gamete possesses all the morphological structures essential for fertilization. However, the testicular spermatozoon undergoes numerous physiological and biochemical changes in both the male and female reproductive tracts. These changes are necessary for the male gamete to complete the early stages of fertilization (recognition and penetration of the zona pellucida, fusion of the nuclei) (Grizard and Jimenez, 1997).

- **The Sertoli cell**

Sertoli cells rest on the basement membrane and are connected to one another by complex junctions. The complex formed by these intercellular junctions and the *tunica propria* constitutes the blood-testis barrier (Dym and Fawcett, 1970).

The Sertoli cell plays several roles, including:

- supporting and supplying nutrients to germ cells (Grootegoed et al., 1989);
- synthesizing and secreting androgen-binding protein (ABP), transferrin for inhibin transport, activin for regulation, plasminogen activator, and ceruloplasmin (Carreau et al., 1994);
- the synthesis and secretion of estrogens following the aromatization of androgens;
- the phagocytosis of dead germ cells and the resorption of residual bodies (Dadoune and Démoulin, 1991).

III-1-2-2- Endocrine function: steroidogenesis

The endocrine function of the testis is carried out by Leydig cells and Sertoli cells. Through these two cell types, the testis primarily synthesizes two types of steroids: androgens and estrogens.

- **Androgens**

Testicular steroidogenesis is primarily Leydig cell-mediated, and the main steroids synthesized are androgens. This synthesis of androgens occurs under the action of LH (luteinizing hormone), starting from cholesterol in the endoplasmic reticulum of Leydig cells. The cholesterol used is primarily drawn from the bloodstream in the form of lipoproteins. It may also originate from local synthesis from acetate (Dadoune and Démoulin, 1991). In the mitochondria, the side chain of cholesterol is cleaved, leading to the formation of pregnenolone, under the action of an enzyme complex containing cytochrome $P\text{-}450_{scc}$ (Robel, 1991). In the microsomes of the endoplasmic reticulum, pregnenolone is converted either directly into testosterone ($\Delta_4$pathway) or into androstenediol and then into testosterone ($\Delta_5$pathway) (Johnson and Everitt, 1988; Hanukoglu, 1992).

- **Estrogens**

The testes also synthesize estrogens (estradiol, estrone, estriol) through the aromatization of androgens (Koueta and Kamtchouing, 1992).

III-1-2-3- Transport, metabolism, and biological actions of androgens

In the blood, approximately 50% of testosterone is transported by a β-globulin, TeBG (testosterone-binding globulin), which has a high affinity. It also transports dihydrotestosterone (DHT) and estradiol-17β. Only the fraction not bound to TeBG is capable of entering target cells. Testosterone can also be transported by non-specific proteins such as albumin and transcortin (Mercier-Bodard et al., 1970). In the testes, synthesized testosterone enters the seminiferous tubules by simple diffusion, passes into Sertoli cells where, after reduction to DHT, it is taken up by ABP (androgen-binding protein). In seminal plasma, androgen transport is mediated by ABP (Drosdowsky and Mahoudeau, 1980).

In target organs such as the seminal vesicles and the prostate, testosterone is converted by the enzyme 5α-reductase into the more active 5α-dihydrotestosterone (Fournier-Delpech and Thibault, 1991). It also serves as a precursor for the formation of estrogens. DHT and 17β-estradiol are the active metabolites of testosterone. In the liver, most testosterone is converted into 17-ketosteroids and androstanediols, which are excreted in the urine. These compounds are inactive.

Androgens, the most important of which is testosterone, are responsible for the differentiation of the internal structures of the male genital system from the Wolffian duct during fetal life and for the masculinization of the external genitalia or the urogenital sinus indirectly via DHT (Jost, 1991). At puberty, androgens are responsible for the development and maintenance of secondary sexual characteristics, the development and secretory activity of the accessory glands, and the initiation and maintenance of spermatogenesis. In the epididymis, testosterone stimulates the synthesis of specific proteins essential for the post-testicular differentiation of spermatozoa (Fournier-Delpech and Thibault, 1991; Lejeune et al., 1994; Saez et al., 1995; Carreau, 1996).

III-1-2-4- Regulation of testicular functions

Each of the two testicular functions is dependent on one of the two gonado -stimulating hormones: FSH and LH. However, numerous findings suggest that testicular activity depends on other endocrine factors (prolactin, GH, insulin) and on numerous molecules known to have hormonal activity, called growth factors.

- **Cerebro-testicular axis (Marieb, 1999)**

The hormonal regulation of spermatogenesis and testicular androgen production involves interactions between the hypothalamus, the anterior pituitary, and the testes. Indeed, the hypothalamus secretes gonadotropin-releasing hormone (Gn-RH, "gonadotropin-releasing hormone"), which regulates the release of gonadotropins—follicle-stimulating hormone (FSH) and luteinizing hormone (LH)—by the anterior pituitary.

FSH indirectly stimulates spermatogenesis in the testes by binding to Sertoli cells and triggering the secretion of ABP (androgen-binding protein) by supporting epithelial cells. ABP allows cells in the convoluted seminiferous tubules to bind testosterone. The ABP-testosterone complex acts on germ cells to promote the continuation of meiosis and spermatogenesis. FSH thus makes the cells receptive to the stimulatory effects of testosterone (Marieb, 1999).

LH binds to interstitial cells (Leydig cells) and stimulates them to secrete testosterone.

The hypothalamus and the anterior pituitary may be subject to the inhibitory action of certain hormones present in the blood. Testosterone inhibits the secretion of gonadotropin-releasing hormone (Gn-RH) by the hypothalamus, and it is thought that it may act directly on the anterior pituitary to inhibit the release of gonadotropins (FSH and LH). Inhibin is a hormone secreted by supporting epithelial cells. When sperm count is high, inhibin secretion increases, which directly inhibits the release of FSH by the anterior pituitary and of Gn-RH by the hypothalamus. When sperm count falls below 20 million per milliliter, inhibin secretion drops sharply and spermatogenesis resumes.

- **Regulation by Other Hormones**

Prolactin (PRL) plays a positive role in steroidogenesis and the endocrine regulation of spermatogenesis, by increasing the number of Leydig cell LH receptors and stimulating the binding of LH to its receptors (Dadoune and Démoulin, 1991).

Growth hormone (GH): In rodents, GH deficiency is associated with delayed puberty or a weak response of Leydig cells to LH/hCG stimulation (Chatelain et al., 1991).

Insulin potentiates the steroidogenic action of LH by stimulating glucose transport, glycogenolysis, and lactate production in Sertoli cells. It also stimulates the production of certain proteins, notably ABP, transferrin, and inhibin (Carreau, 1996; Aquila et al., 2004).

Glucocorticoids inhibit steroidogenesis by reducing the activity of 17α-hydroxylase in Leydig cells (Huhtaniemi and Toppari, 1996).

III-2- The Epididymis and Accessory Glands

III-2-1- The epididymis (Marieb, 1999)

- **Structure**

The epididymis is a tube attached to the testis. It consists of three regions: the head, the body, and the tail, which often contains a significant reserve of spermatozoa. The epididymal duct is highly coiled and composed of lobules separated by connective tissue septa visible in the head and tail.

- **Functions**

The epididymis ensures the transport, survival, and functional maturation of sperm.

The epididymis synthesizes substances that support sperm metabolism. This includes acetylcarnitine and inositol, which play a role in sperm survival, differentiation, and maturation. It also includes α-glucosidase, which catalyzes the hydrolysis of polysaccharides directly into glucose. The sperm's increasing motility and their ability to attach to the zona pellucida are acquired throughout their passage through the epididymis.

III-2-2- Accessory Glands

The development and secretory activity of the accessory glands are androgen-dependent. Their secretions constitute approximately three-quarters of the seminal plasma in an ejaculate (Setchell , and Brooks, 1988).

- **The seminal vesicles**

They consist of two multilobed sacs whose epithelium may be deeply folded, opening either into the terminal portion of the vas deferens or into the urethra. Their secretion, which accounts for approximately 60% of the semen volume, is an alkaline, viscous, yellowish fluid. In humans, their secretion consists mainly of fructose, but also contains ascorbic acid, prostaglandins, and insulin (Stahler et al., 1987).

In rats and mice, the seminal vesicles are attached to the coagulating glands. Fructose is synthesized primarily by the coagulating glands, while the seminal vesicles primarily synthesize citric acid. Seminal fructose is derived from glucose metabolism via the polyol pathway, as the coagulating glands contain the necessary enzymatic machinery, namely aldose reductase and sorbitol dehydrogenase (ketose reductase) (Samuels and Harding, 1962; Prendergast and Veneziale, 1975).

Fructose is essential for the survival and motility of spermatozoa (Prendergast and Veneziale, 1975).

- **The Prostate**

It is a gland that is most often attached to the urethra. In humans, prostatic secretion is rich in zinc and spermine. Prostatic fluid has an acidic pH and is very rich in enzymes, particularly acid phosphatases and proteases. However, it also contains other cations such as Ca^{2+} , Mg^{2+} , and K^{+} . Zinc in seminal plasma is thought to confer fertilizing capacity to germ cells. This is believed to help stabilize the nucleoproteins of sperm DNA (Fournier-Delpech and Thibault, 1991).

In mice, the prostate secretes fructose and citric acid (Samuels and Harding, 1962).

- **The bulbourethral glands**

Also known as Cowper's glands, they are located beneath the prostate and secrete a thick, translucent mucus that flows into the spongy portion of the urethra. This secretion is released prior to ejaculation and serves to neutralize the acidity of any traces of urine still present in the urethra (Marieb, 1999).

IV- GENERAL INFORMATION ON THE LIVER

IV-1- Anatomy

Located in the upper right quadrant of the abdomen beneath the diaphragmatic dome, the liver is a solid, brownish-colored organ with a smooth, homogeneous surface surrounded by a connective tissue capsule: Glisson's capsule. This capsule invaginates, forming several furrows that delineate four lobes, including two main lobes separated on the surface by the falciform ligament: the right lobe, which is the largest, and the left lobe, to which the caudate lobe and the square lobe are attached. The gallbladder is located in a small fossa on the underside of the liver, at the level of the right lobe (Bourel, 1991). Connected to the digestive tract, the liver is the largest and densest organ in the body. In humans, it weighs approximately 2 kg, of which 800 to 900 g is blood (Dadoune et al., 1990). The liver has a dual vascular system: afferent (arterial and portal) and efferent. The afferent arterial system supplies oxygen and metabolites to the cells (nutritional circulation); the portal circulation drains all venous blood from the digestive tract (functional circulation). Efferent circulation is provided by the suprahepatic veins, which drain blood from the liver to the inferior vena cava (Bourel, 1991).

IV-2- Histology

The hepatic parenchyma consists of lobules, each centered by a terminal suprahepatic venule or centrolobular vein. Each angle of the lobule is formed by a portal space containing a branch of the hepatic artery, the portal vein, the bile ducts, and the lymphatic capillaries. The lobule is the structural unit of the parenchyma, whereas the simple acinus corresponds to the smallest functional unit of the liver. The acinus is a portion of the parenchyma supplied by the same set of branches of the hepatic artery and vein originating from two adjacent portal spaces (Dadoune et al., 1990). Between the angles and the center of the lobule, the hepatocyte rows (Remak's rows) are arranged radially, separated by sinusoids that converge toward the central vein (Di Fiore, 1965).

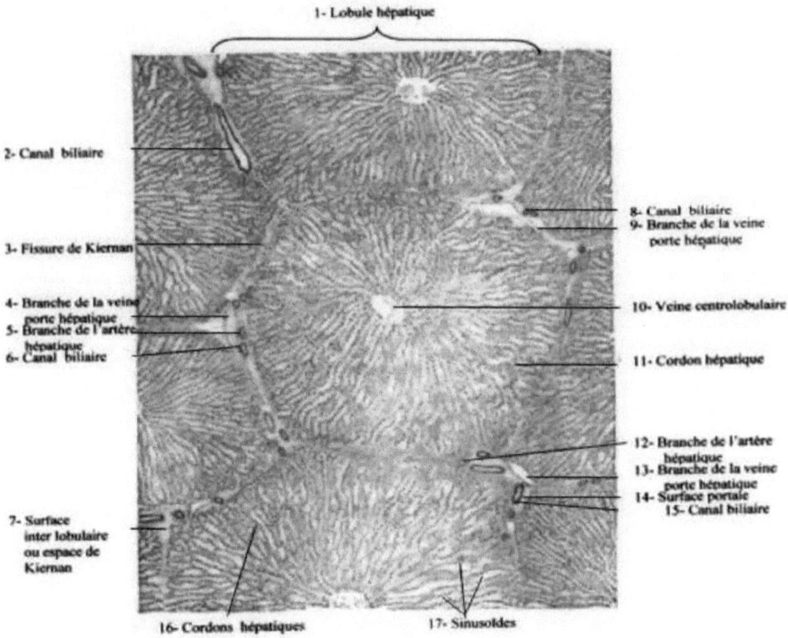

Figure 5: Histological section of a hepatic lobule (panoramic view) ×45 (Di Fiore, 1965)

IV-3- Functions

The liver performs various functions:

- defense of the body: this defense is carried out by Kupffer cells, which are a type of large white blood cell whose purpose is to clear the blood of debris such as bacteria and spent red blood cells (Bourel, 1991);

- bile production: secreted by hepatocytes, bile is necessary for digesting fats in the intestine (Guyton and Hall, 1996);

- carbohydrate metabolism: carbohydrates, broken down into sugar molecules in the small intestine, are converted into glycogen stored in the liver under the influence of the hormone insulin. When blood glucose levels drop, glycogen in the liver is converted into glucose under the influence of the hormones adrenaline and glucagon (Guyton and Hall, 1996);

- lipid metabolism: in the liver, fats are captured in the sinusoidal space of Disse and metabolized either for energy production (Krebs cycle) or for the synthesis of fatty acids, triglycerides, phospholipids, cholesterol, and lipoproteins (Guyton and Hall, 1996);

- blood detoxification: the liver inactivates substances that are toxic to the body. Active substances, such as many medications, combine with oxygen, rendering them harmless. It removes ammonia from the blood, which is converted into urea and excreted in the urine (Guyton and Hall, 1996);

- protein metabolism: the liver plays a role in the breakdown of peptides and their amino acids of intestinal origin. These peptides lose their amino acids and are converted into urea. It synthesizes many structural and functional proteins such as: albumin, fibrinogen, globulin, transferrin, coagulants... (Guyton and Hall, 1996).

MATERIALS AND METHODS

1- EXPERIMENT

I-1- The Animals

The animals used in our experiment were white BALB C strain *Mus musculus* mice. Some of these animals came from our animal facility, where they were raised. The others, which came from the Pasteur Center, were acclimated under normal conditions (room temperature, adequate ventilation, natural lighting). Their diet consisted of corn flour (47%), wheat flour (15%), fish meal (15%), soybean meal (20%), palm kernel meal (1.1%), bone meal (1%), table salt (0.8%), and a vitamin complex (0.1%) (Vitamino) added to the drinking water. The drinking water was tap water, and all animals were fed *ad libitum*.

I-2- Plant material and preparation of the aqueous extract of *Clerodendrum umbellatum*

The leaves of *Clerodendrum umbellatum* were harvested in April 2005 in Djoum (Nkolenyen) in the South Province (Cameroon). These fresh leaves were air-dried and finely ground using a grinder, yielding a powder.

The aqueous extract of *Clerodendrum umbellatum* leaves was prepared using 300 g of *Clerodendrum umbellatum* leaf powder macerated in 3 liters of water. The mixture was stirred several times. After 24 hours, this mixture was filtered using Whatman No. 3 filter paper. The filtrate was then evaporated in an oven at 45°C. This process yielded 63.62 g of crude extract, representing a yield of 21.20%.

I-3-Phytochemical tests of the aqueous extract of *Clerodendrum umbellatum* leaves (Trease and Evans, 1983)

These tests were conducted to detect the presence or absence of certain chemical constituents in the extract.

- **Triterpene test (Liebermann-Burchard reaction)**

The mixture of 2 mL of extract solution (1.5 g of extract in 15 mL of distilled water) with the reagent consisting of 3 µL of concentrated sulfuric acid, 3 µL of acetic anhydride, and 3 µL of chloroform produced a bluish-green color in the case of steroids and a purplish-red color in the case of triterpenes.

- **Saponin test**

5 mL of the extract solution was vigorously shaken. The formation of a persistent foam lasting longer than 10 minutes indicated the presence of saponins in the extract.

- **Saponoside test**

The presence of persistent foam during aqueous extraction or after shaking the extract solution, along with a positive reaction to the Liebermann-Burchard test, confirmed the presence of saponosides in the sample.

- **Tannin test**

1 1 mL of extract solution in the presence of 2 µL of ferric chloride ($FeCl_3$) yields a blackish-green color for catechin tannins and a dark blue color for gallic tannins.

- **Flavonoid test**

To 1 mL of the extract solution, 0.5 mL of ethyl alcohol was added. 3 µL of concentrated hydrochloric acid (HCl) and a few magnesium shavings were then added. The presence of flavonoids is indicated by a change in the solution's color (effervescence and a color change that may be violet or brick red).

- **Alkaloid test**

1 1 mL of the extract solution was mixed with 3 µL of concentrated hydrochloric acid and 2 µL of Meyer's reagent. The formation of a precipitate indicates the presence of alkaloids.

- **Phenol test**

1 mL of extract solution was mixed with 2 µL of ferric chloride ($FeCl_3$). The presence of phenols is indicated by the formation of a blue or violet complex (color change).

- **Cardiac glycoside test (Salkowski test)**

2 mL of extract solution was diluted in approximately 2 mL of chloroform. After shaking, a few drops of sulfuric acid (H_2SO_4) solution were carefully added to produce two phases. The formation of a reddish-brown zone between the two phases indicates the presence of glycosides.

- **Lipid test**

A few drops of plant extract solution were placed on filter paper. The paper was then dried.

The observation of translucent spots at the sites where the drops were placed indicates the presence of lipids.

I-4- Collection of mollusks

The intermediate hosts of the schistosomes were mollusks of the species *Biomphalaria pfeifferi*. They were collected in the city of Yaoundé at Municipal Lake by systematically combing the shoreline vegetation using a fine-mesh net with a handle approximately one meter long. They were removed from the net using tongs and placed in plastic jars containing lake water. To protect against potential infestations, wearing a pair of waterproof boots and gloves was necessary. The mollusks brought back to the laboratory were carefully cleaned and placed in plastic containers containing rainwater and lettuce for food.

I-5- Infestation of mice (definitive hosts)

After counting, the *Biomphalaria pfeifferi* snails were distributed into crystallizers and exposed to a light source from a 40 W lamp to accelerate cercariae release. The egg-laying snails were identified, counted, and used to infest the mice. Out of a total of 3,283 snails collected from 8 malacological surveys, 88 were identified as laying *Schistosoma mansoni* cercariae, representing a natural infestation rate of 2.68%.

Male white mice, approximately 90 days old and weighing between 23 and 32 g, were divided into 7 groups as shown in Table 2. Mice in all groups were first weighed and placed in cages with rainwater to stimulate defecation and urination. They were then transferred individually to containers containing rainwater. Fifty freshly released cercariae from at least three mollusks, collected under a binocular microscope (ZEISS Stemi DV4), were introduced into the containers containing the mice. Exposure to the cercariae lasted 60 minutes.

Table 2: Distribution of mice by group and number of cercariae administered

Groups	**Number**	**Dose of cercariae administered**
Tn	10	0
Tp60	14	50
Tp81	28	50
E40	14	50
E80	14	50
E160	14	50
E320	14	50

Tn: negative controls, Tp60: untreated infested controls sacrificed on the 60th day after infestation, Tp81: untreated infested controls sacrificed on the 81st day after infestation; E40, E80, E160, E320: test groups infested and then treated with doses of 40, 80, 160, and 320 mg/kg of the aqueous extract of *C. umbellatum* leaves, respectively.

Mice in the Tn group (negative control) underwent the same stress as the others, with the sole difference that they were not infested. After infestation, the animals were returned to breeding cages and fed *ad libitum*. The animals' weight gain was assessed weekly throughout the experimental period (81 days).

I-6- Determination of ovarian burden and treatment of infested animals

I-6-1- Determination of ovarian burden by the Kato-Katz test (WHO, 1994)

To confirm the infestation, a stool examination was performed before treatment began on the^{60th}day after infestation (^{8th}week) to identify *Schistosoma mansoni* eggs in the feces. The Kato-Katz technique (WHO, 1994) was used for this purpose. It involves placing a small amount of stool on a piece of paper and pressing a nylon sieve onto the sample so that some of the fecal matter passes through the sieve. Using a flat-edged spatula,

the top surface of the mesh was scraped to collect the feces, which were used to fill the hole in the perforated plate (template) placed in the center of a microscope slide. Once the excess stool was removed, the plate was removed and the fecal matter covered with a rectangle of cellophane soaked in a malachite green glycerol solution (100 mL glycerol, 5 mL malachite green, and 100 mL distilled water). The stool was smeared onto the slide by rolling a hemolysis tube over the cellophane. The preparations were examined under a ZEISS Standard 25 optical microscope (100x magnification). The number of *Schistosoma mansoni* eggs counted was multiplied by 24 (slide with a 6-mm-diameter hole and 1.5-mm thickness) to obtain the number of eggs per gram of stool.

At the end of this examination, it was found that of the 98 mice initially infested, 81 were parasitized, representing an infestation rate of 82.6%.

I-6-2- Treatment of the animals

- **Preparation of the various solutions**

The solvent used to dissolve the extract was distilled water.

The volume of extract administered to each mouse was determined using the following formula:

$$V = \frac{P \times D}{C}$$

Where P is the weight of the mouse in kg,

D is the dose of extract administered in mg/kg,

C is the concentration of the solution in mg/mL

- **Administration of the extract**

Administration of the aqueous extract of *Clerodendrum umbellatum* began on the 60^{th}day after the mice were infected with *Schistosoma mansoni* cercariae. It was completed on the 74thday. The extract doses (40, 80, 160, and 320 mg/kg) were administered orally via an esophageal tube daily for 14 days (Table 3). Mice in the Tn and Tp81 groups received distilled water.

Table 3: Distribution of mice by group and dose of extract administered

Groups	Number of mice		Extract doses (mg/kg)
	Before treatment	After treatment	
Tn	10	10	0
Tp60	10	10	0
Tp81	23	10	0
E40	10	10	40
E80	13	11	80
E160	13	12	160
E320	12	11	320

Tn: negative controls; Tp60: untreated infested controls sacrificed on the 60^{th}day after infestation; Tp81: untreated infested controls sacrificed on the 81^{st}day after infestation; E40, E80, E160, E320: test groups infested and then treated with doses of 40, 80, 160, and 320 mg/kg, respectively, of the aqueous extract of *Clerodendrum umbellatum* leaves.

The animals in the Tp60 group were sacrificed on the 60^{th}day. At the end of treatment, the animals in the group were kept under observation for 7 days. At the end of these 7 days, it was found that of the 48 animals that began treatment on the 60^{th}day, 4 deaths were recorded during the treatment period, and 44 survived to the 81^{st}day, representing a mortality rate of 8.33%.

Regarding the untreated parasitized animals sacrificed on the 81^{st}day post-infestation, out of 23 animals gathered on the 60^{th}day, 13 deaths were recorded, and only 10 reached the 81^{st}day, resulting in a mortality rate of 56.52%.

I-7- Sample Composition

At the end of the 81-day experimental period (60 days post-infestation, 14 days of treatment, and 7 days of observation), a portion of the animals in each group was sacrificed by decapitation, and the remainder were perfused to isolate the schistosomes. Prior to sacrifice, feces from each mouse were collected over a 24-hour period to count *S. mansoni* eggs.

I-7-1. Blood and Organ Collection

In mice sacrificed by decapitation, arterial-venous blood was collected in EDTA tubes and centrifuged (Eppendorf centrifuge) at 3,500 rpm for 15 minutes. The resulting plasma was distributed into Eppendorf tubes and stored at -20°C for biochemical analysis.

The liver, spleen, testes, epididymis, and seminal vesicles were removed and weighed immediately after sacrifice. A portion of the liver was preserved in 10% formalin for histopathological analysis. The remaining portion of the liver was weighed and then digested in 4% KOH for counting *S. mansoni* eggs. The left testis, right epididymis, and seminal vesicles were stored at -20°C for the determination of testosterone, α-glucosidase, and seminal fructose levels. The tail of the left epididymis was dissected in Earle's solution for sperm counting.

The relative weight (organ weight per 100 g of body weight) of all harvested organs was determined using the following formula:

$$PR = \frac{\text{Poids de l'organe}}{\text{Poids corporel}} \times 100$$

I-7-2- Recovery of schistosomes

In mice perfused for schistosome recovery, the technique of Duvall & Dewitt (1967) was used.

The mice were euthanized by intraperitoneal injection of a lethal dose of anesthetic (0.5 mL of 6% sodium pentobarbital). This not only causes rapid death of the mouse but also paralyzes the schistosomes, which then detach from the endothelium of the mesenteric veins and can thus be carried by the bloodstream to the liver.

The dead mouse was suspended by its left front and hind legs from the vertical support of the perfusion chamber. The abdomen and thorax were incised to expose the intestine, liver, and heart. The portal vein was transected just beyond the confluence of the mesenteric veins. Immediately, a hypodermic needle was inserted into the left ventricle, and the perfusion fluid (8.5 g of sodium chloride and 7.5 g of sodium citrate in 1 L of distilled water) was injected. The schistosomes were thus carried by the perfusion fluid toward the severed portal vein and collected in a beaker placed beneath the rodent. The perfusion was continued until the liver and mesenteric veins were completely decolorized. The viscera were rinsed with a spray from a syringe filled with physiological saline (9‰ NaCl) to recover any worms that might be trapped between the viscera.

After perfusion, the schistosomes were collected by filtering the contents of the beaker through a 25-µm sieve. The rodent's viscera were removed and examined under a magnifying glass to recover any schistosomes that might have remained trapped (Tchuem Tchuente, 1993). The schistosomes collected in this manner were then counted.

The percentage reduction in schistosomes following treatment with the aqueous extract of *Clerodendrum umbellatum* leaves was calculated according to the method of Tendler et al. (1986), as follows:

$$\mathbf{P = [(C-V)/C] \times 100}$$

P: Percentage reduction in worms or protection provided by the extract

C: Average number of parasites recovered from untreated infected animals **V**: Average number of parasites recovered from treated animals

After schistosomes were recovered, the liver was removed and weighed; a small portion of known weight was digested in 4% KOH for counting *S. mansoni* eggs.

I-7-3- Counting of eggs in the liver

A fragment of the liver from each mouse was weighed and placed in an Erlenmeyer flask containing 100 mL of a 4% potassium hydroxide solution (Cheever's method, 1968). The flask was then placed in a water bath at 37°C for 5 to 7 hours. The organ tissues were thus dissolved, and the intact eggs were released into the solution (Tchuem Tchuente, 1993). The resulting homogenate was centrifuged (Eppendorf centrifuge) at 1,500 rpm for 5 minutes. The pellet was mounted between a slide and a coverslip, and the preparation was examined under an optical microscope (ZEISS Standard 25) at 100× magnification. The number of *S. mansoni* eggs was counted and then estimated per gram of liver.

I-7-4- Sperm Counting in the Epididymis

The physiological solution for determining the concentration of epididymal spermatozoa was prepared by dissolving 0.884 g of Earle's powder; 0.008 g of penicillin; 0.001 g of sodium pyruvate; and 0.021 g of sodium bicarbonate in 100 mL of distilled water.

The tail of the left epididymis of mice sacrificed by decapitation was dissected into a Petri dish containing 1 mL of the physiological solution. Next, a drop of the supernatant (4 µL) was placed in the center of the chamber of a hemocytometer (Malassez chamber), and the sperm concentration was determined by counting in 10 of the 25 squares of the chamber using a ZEISS Standard 25 optical microscope (400× magnification). The number obtained was multiplied by 1,000,000 (WHO, 1993).

I-7-5- Preparation of homogenates

- **Preparation of potassium phosphate buffer (0.1 M; pH 6.8)**

The potassium phosphate buffer was prepared by separately dissolving 6.80 g of KH_2PO_4 and 8.70 g of K_2HPO_4 in 500 mL of distilled water. The pH of the first solution was adjusted to 6.8 with the second solution. The final solution was stored at 4°C and was stable for one month.

- **Epididymis**

The right epididymis of each mouse was weighed, minced, and homogenized to 20% in potassium phosphate buffer (0.1 M; pH 6.8) using a Teflon Potter on an ice bath. After centrifugation (Eppendorf centrifuge) at 10,000 g for 30 minutes at 4°C, the recovered supernatant was stored at -20°C for α-glucosidase assay.

- **Seminal vesicles**

On an ice bath, the seminal vesicles from each mouse were crushed and homogenized with 25% distilled water. After centrifugation (Eppendorf centrifuge) at 3000 g for 15 minutes at 4°C, the recovered supernatant was stored at -20°C for fructose measurement.

II- BIOCHEMICAL ANALYSES

II-1- Measurement of epididymal acid α-glucosidase activity (WHO, 1993)

- **Principle**

Paranitrophenyl-α-D-glucopyranoside (PNPG) in the presence of α-glucosidase is converted to paranitrophenol (PNP) and glucose at pH 3.8. The paranitrophenol formed turns yellow in the presence of the stop reagent (0.1 M glycine/NaOH buffer; pH 10.6), and the optical density is measured using a spectrophotometer (SECOMAM CE BASIC 70VB358) at 405 nm.

- **Reagents**

- **0.1 M phosphate/citrate buffer; pH 3.8**

Two solutions, A and B, were prepared by dissolving 1.36 g of KH_2PO_4 in 100 mL of distilled water and 2.10 g of citric acid ($C_6H_8O_7$, H_2O) in 100 mL of distilled water, respectively. The pH of solution A was adjusted to 3.8 using solution B, and the mixture was stored at 4°C.

- **0.1 M glycine/NaOH buffer; pH 10.6 (stop reagent)**

7.5 g of glycine were dissolved in 300 mL of distilled water, and the pH was adjusted to 10.6 using a NaOH solution. The final solution was made up to 500 mL with distilled water and stored at 4°C.

- **PNP standard solution (0.1 mM)**

The 0.1 mM PNP standard solution was prepared by diluting 0.1 mL of 0.01 M PNP stock solution in 9.9 mL of distilled water.

- **Procedure**

- **Calibration**

The PNP standard curve was established prior to sample analysis.

Table 4: α-Glucosidase Assay (Calibration)

Standard tubes	I	II	III IV	V	VI
0.1 mM PNP (µL)	0	150	300	450	600
Distilled water (µL)	750				
Final volume (µL)	750	600	450	300	150
	0				
	750	750	750	750	750

	750				
Mix and incubate at 37°C for one hour					
Stop reagent (mL) Number of PNP moles (moles)	2.5 2.5 0 0.075	2.5 0.015	2.5 0.030	2.5 0.045	2.5 0.060

The measurement of α-glucosidase activity in epididymal homogenates was performed as shown in Table 5

Table 5: Measurement of epididymal α-glucosidase activity

	Blank	Samples
Distilled water (µL) Epididymal homogenates (µL) Phosphate/citrate buffer (µL) PNPG (µL)	50 - - 100	- 50 600 100
Mix and incubate at 37°C for one hour		
Stop reagent (mL)	2.5	2.5

The optical density of each sample was measured against the blank at 405 nm using a spectrophotometer (SECOMAM CE BASIC 70VB358).

- **Calculation of α-glucosidase activity**

One unit of α-glucosidase corresponds to the production of one micromole of PNP per minute at 37°C. The amount of PNP in each sample was determined based on the measured optical density. Since the incubation time was 60 minutes, the activity was calculated by dividing the amount of PNP obtained by 60. The assay volume was 50 µL. This activity corresponds to the amount of protein contained in this volume. The activity was expressed in mU/g of organ.

II-2- Seminal fructose assay (WHO, 1993)

- **Principle**

In an acidic medium, indole reacts with fructose in solution to produce a brown color. The intensity of the color is proportional to the concentration of fructose in the medium.

- **Reagents**

- **1.8% $ZnSO_4$, 7H2O solution**

This solution was prepared by dissolving 1.8 g of $ZnSO_4 \cdot 7H_2O$ in 100 mL of distilled water. It was stored at room temperature.

- **0.1 M NaOH solution**

This solution was prepared by dissolving 0.2 g of NaOH in 50 mL of distilled water. It was stored at room temperature.

- **Indole reagent**

200 mg of benzoic acid were dissolved in 100 mL of distilled water by repeated stirring in a water bath at 60°C. After complete dissolution, 25 mg of indole were added to the solution, and the resulting mixture was filtered using Whatman $^{No.}$ 3 filter paper and stored at 4°C·

- **Fructose stock solution (2.8 mM)**

The fructose stock solution was prepared by dissolving 25.2 mg of fructose in 50 mL of distilled water. Part of this solution was used immediately for the assay, and the remaining portion was stored at -20°C for subsequent assays. The standard solutions used were prepared by diluting the stock solution to concentrations of 0.14 mM and 0.28 mM.

- **Procedure**

- **Sample preparation**

Each sample (0.1 mL) was diluted in 4.9 mL of distilled water. To 1 mL of the diluted sample, 0.3 mL of zinc sulfate (1.8%) and 0.2 mL of NaOH (0.1 M) were successively added. After mixing, the tubes were allowed to stand, then centrifuged for 20 minutes at 2000 g. The supernatant was used for the assay.

- **Assay**

The various reagents were added as shown in Table 6.

Table 6: Determination of fructose concentration using the indole colorimetric method

10 mL glass tubes	Blank	Standard 1	Standard 2	Sample
0.14 mM fructose (mL)	-	0.5	-	-
Fructose 0.28 mM (mL)	-	-	0.5	-
The diluted sample and deproteinized [supernatant] (mL)	-	-	-	0.5
	0.5	-	-	-
Distilled water (mL)	0.5	0.5	0.5	0.5
Indole (mL)	5	5	5	5
Concentrated HCl (mL)				

All tubes were then capped and incubated in a water bath at 50°C for 20 minutes. After rapid cooling in melting ice to room temperature, the optical density was measured using a spectrophotometer (SECOMAM CE BASIC 70VB358) at 470 nm against a blank.

▪ **Calculation of fructose concentration**

The fructose concentration (µmol/g of tissue) in the samples was calculated using the following formula: **As x F x f**

Where As: optical density of the sample;

f: dilution factor (75);

F: mean value of the fructose standard.

$$F = \frac{1}{2}\,[0.14 / s_1 + 0.28 / s_2]$$

s_1: average optical density measured for 0.14 mM fructose

s_2: average optical density measured for 0.28 mM fructose

The fructose concentration was determined by taking into account the homogenization percentage of the organs, the dilution factor, and the weight of the seminal vesicles.

III- HISTOLOGICAL STUDY

III-1- Preparation of solutions

- **Meyer's hemalun**

6 g of potassium alum were dissolved in 300 mL of distilled water to obtain a 2% potassium alum solution. This solution was brought to a boil. 0.6 g of hematine powder were then added, and the mixture was kept boiling for 5 minutes. After cooling, 6 mL of glacial acetic acid were added, and the mixture was filtered before use.

- **1% acidic water solution**

250 mL of this solution was prepared by adding 247.5 mL of distilled water to a beaker containing 2.5 mL of acetic acid.

- **Lithium hydroxide solution**

The solution is prepared by placing 300 mL of distilled water in a beaker; while stirring with a magnetic stirrer, lithium carbonate is gradually added until saturation is reached.

- **Methylene blue solution**

2 g of methylene blue were dissolved in 100 mL of distilled water. After complete dissolution,

2.5 mL of acetic acid were added, and the solution was filtered before use.

- **Gurr® R:11 S:7-16 trichrome solution**

This solution was ready for use.

III-2- Procedure

III-2-1- Dehydration, embedding, and mounting

After fixation in 10% formalin, the liver was cut (transversely) into small slices and placed into cassettes numbered with a pencil. Dehydration was performed by placing the cassettes in baths of ethyl alcohol of increasing concentration: 80° (for 15 minutes), 95° (for 30 minutes), 100° (for 30 minutes), and 100°(for 30 minutes). Upon removal from the final 100° alcohol bath, the organ was clarified by passing the cassettes through three toluene baths at intervals of 15 minutes, 30 , and 30 minutes, respectively. The actual embedding consisted of placing the cassettes in a paraffin dispenser containing paraffin melted at 60°C for 2 hours. Next, the samples were placed in a mold into which we poured paraffin. After cooling, we obtained blocks. The blocks thus prepared were placed in a freezer for further hardening.

III-2-2- Preparation of sections

The blocks removed from the freezer were mounted on the microtome (Leitz Wetzlar) and cut to a thickness of 5 µm. This yielded a paraffin strip from which various sections were selected and placed on a slide that had previously been coated with glycerinated albumin solution. The slide was then placed on a heating plate (45°C) to smooth out wrinkles and dried in an oven (37°C) for 15 minutes.

III-2-3- Staining

The slides removed from the oven were deparaffinized in three tanks of toluene for 5 minutes each, then rehydrated by successive passes through 100% ethanol (3 minutes), 100% ethanol (3 minutes), 95% ethanol (3 minutes), and 80% ethanol (2 minutes). After rinsing with tap water, the liver slides were stained in Meyer's hemalun for 5 minutes (nuclear staining) and then rinsed with water until the stain was removed. They were then instantly rosy-stained by passing them through acid water. After rinsing with tap water, they were blue-stained in lithin water, then rinsed again. After verifying under the microscope that the nuclei were blue against a clear background, the slides were stained in trichrome solution for 7 minutes. After removing excess stain by rinsing, they were finally stained in a methylene blue solution (7 minutes) and rapidly dehydrated in two trays of absolute alcohol, then soaked in toluene before being mounted with Eukitt.

III-2-4- Observation and Photography

The stained sections were observed under an Olympus CH2 photonic microscope, and photographs were taken using an Olympus 101 camera mounted on the microscope.

IV- STATISTICAL ANALYSIS OF RESULTS

The results were expressed as mean ± SEM (standard error of the mean). Statistical analysis of the results was performed using StatsDirect software. Analysis of variance (ANOVA) followed by the nonparametric Kruskal-Wallis test was used to compare the results of the different treatment and control groups.

RESULTS

I- RESULTS OF PHYTOCHEMICAL TESTS OF THE AQUEOUS EXTRACT OF *CLERODENDRUM UMBELLATUM* LEAVES

The results of phytochemical tests indicate that the aqueous extract of *Clerodendrum umbellatum* leaves contains alkaloids, flavonoids, saponins, saponosides, tannins, and triterpenes. Cardiac glycosides, lipids, steroids, and phenols, however, are absent.

Table 7: Phytochemistry of the aqueous extract of *Clerodendrum umbellatum* leaves

Compounds tested	Present (+) or absent (-)
Alkaloids	+
Flavonoids	+
Cardiac glycosides	-
Lipids	-
Phenols	-
Saponins	+
Saponins	+
Tannins	+
Triterpenes	+
Steroids	-

+ : present - : absent

II- WEIGHT CHANGES

Figure 6 shows the weight changes in the mice during the 11-week experiment. This figure indicates that, in general, body weight increases gradually in all infested animals from week 0 to week 7. By week 8, a significant decrease ($p < 0.05$) is observed compared to week 7.

After treatment, the effect of the extract on body weight is mixed; there is an increase in some groups and a decrease in others.

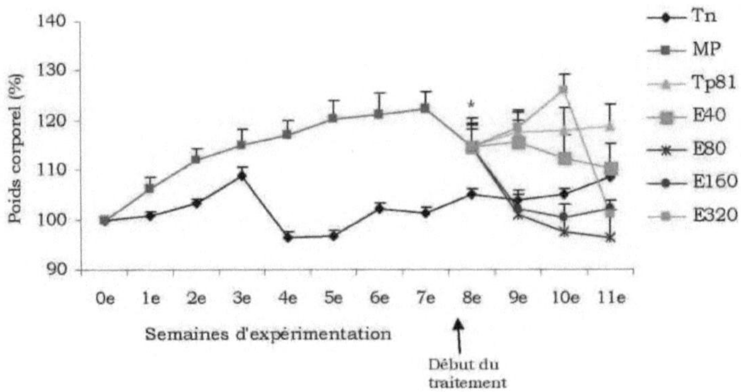

Figure 6: **Body weight changes in mice throughout the experimental period**

Values represent means ± SEM (standard error of the mean)

* $(p < 0.05)$, significant difference compared to week 7

n = group size

Tn: negative control animals (n=10), Tp60: untreated parasitized animals sacrificed on the60thday (n=10), Tp81: untreated parasitized animals sacrificed on the81stday (n=10), E40: animals treated with a dose of 40 mg/kg (n=10), E80: animals treated with a dose of 80 mg/kg (n=11), E160: animals treated with a dose of 160 mg/kg (n=12), and E320: animals treated with a dose of 320 mg/kg (n=11).

III- AVERAGE LOAD OF *S. MANSONI* EGGS IN FECES AND IN THE LIVER

III-1- In feces

Figure 7 (a) shows the presence of *S. mansoni* eggs in feces on the60thday after infestation. After treatment with the aqueous extract of *C. umbellatum* leaves, the number of eggs per gram of feces increased non-significantly in untreated parasitized animals sacrificed on the 81stday, as well as in animals treated with the aqueous extract of *Clerodendrum umbellatum* leaves at doses of 40 and 320 mg/kg. Furthermore, a decrease in the number of eggs was observed in animals treated with doses of 80 and 160 mg/kg, with an overall percentage decrease of 54.48% compared to untreated parasitized animals sacrificed on the81stday (Figure 7(b)).

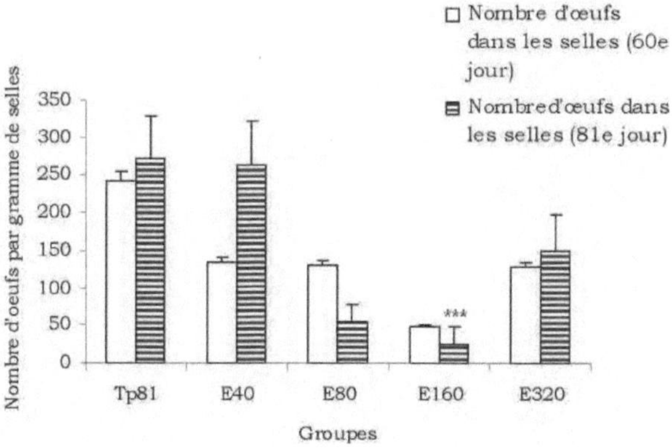

a- Egg shedding in feces between the 60th and 81st day

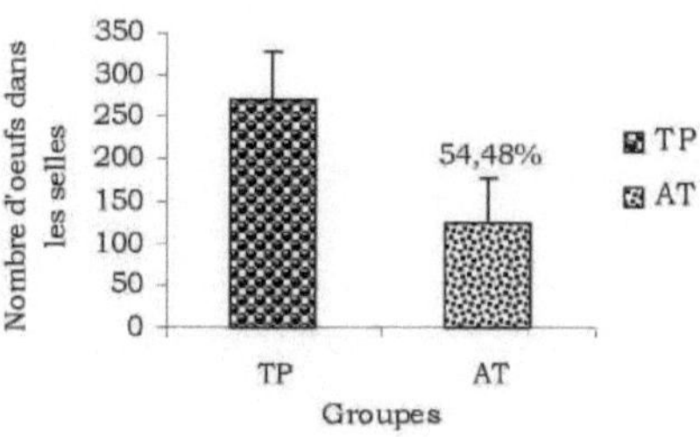

b- Impact of treatment on the egg load in feces

Figure 7: **Effects of the aqueous extract of *Clerodendrum umbellatum* leaves on egg shedding in feces between the 60th and 81st day**

Values represent means ± SEM (standard error of the mean)

*** p<0.001 significant compared to the number of eggs before treatment (60th)

Tp81: untreated parasitized animals sacrificed on the 81st day (n=10), E40: animals treated with a dose of 40 mg/kg (n=10), E80: animals treated with a dose of 80 mg/kg (n=11), E160: animals treated with a dose of 160 mg/kg (n=12), and E320: animals treated with a dose of 320 mg/kg (n=11).

TP: untreated infected controls sacrificed on the 81st day (n=10), AT: treated animals (44).

III-2- In the liver

Figure 8 shows the presence of *S. mansoni* eggs in the livers of untreated infected mice (Tp60 and Tp81). After treatment with the aqueous extract of *C. umbellatum* leaves, a significant decrease (p<0.05 and p<0.001) in the number of eggs in the liver was observed in animals treated at doses of 80 and 160 mg/kg compared to untreated parasitized animals ( ) sacrificed on the 81[th]day (Tp81). Thus, the number of eggs in the liver in all test animals combined after treatment decreased by 22.74% compared to untreated infected animals.

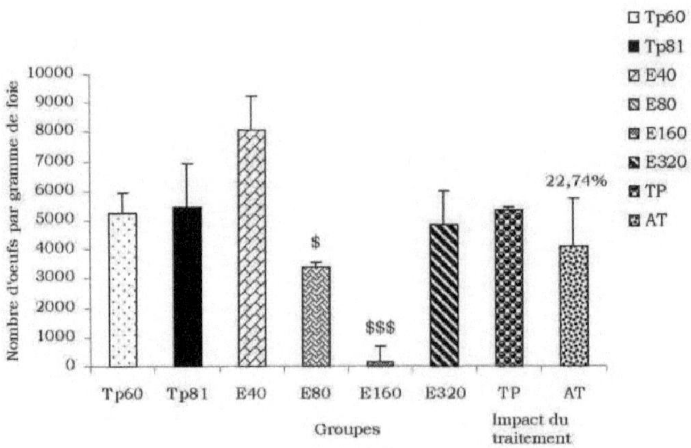

Figure 8: Effects of the aqueous extract of *Clerodendrum umbellatum* leaves on the egg load in the liver

Values represent means ± SEM (standard error of the mean)

$ ($p<0.05$), $$$ ($p<0.001$) significant difference compared to untreated parasitized animals sacrificed on the[81st]day

n = group size

Tp60: untreated parasitized animals sacrificed on the[60th]day (n=10), Tp81: untreated parasitized animals sacrificed on the[81st]day (n=10), E40: animals treated at a dose of 40 mg/kg (n=10), E80: animals treated with a dose of 80 mg/kg (n=11), E160: animals treated with a dose of 160 mg/kg (n=12), and E320: animals treated with a dose of 320 mg/kg (n=11), TP: untreated parasitized controls sacrificed on the 60th[e] and[81st] days (n=20), AT: treated animals (n=44).

V V- AVERAGE SCHISTOSOME LOAD

Figure 9 shows that the number of schistosomes decreased in untreated, parasitized animals sacrificed on the[81st]day compared to those sacrificed on the[60th]day; however, this decrease is not statistically significant. Compared to untreated, parasitized animals sacrificed on the 81st[th]day, a decrease in the number of schistosomes was generally observed in all treated animals, a decrease that was significant ($p<0.01$) at the dose of 160 mg/kg. Thus, the decrease in the average number of schistosomes in all test animals after treatment was 88.45 % compared to untreated parasitized animals.

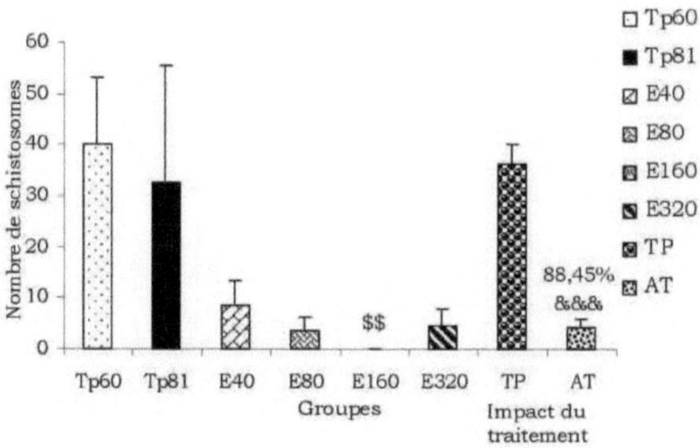

Figure 9: Effects of the aqueous extract of *Clerodendrum umbellatum* leaves on parasite load

Values represent means ± SEM (standard error of the mean)

$$ p<0.01 significant difference compared to untreated parasitized animals sacrificed on the ^{81st}day (Tp81 group)

&&& p<0.001 significant difference compared to untreated parasitized controls sacrificed on the ^{60th}and^{81st}days

n = group size

Tp60: untreated parasitized animals sacrificed on the ^{60th}day (n=10), Tp81: untreated parasitized animals sacrificed on the ^{81st}day (n=10), E40: animals treated at a dose of 40 mg/kg (n=10), E80: animals treated with a dose of 80 mg/kg (n=11), E160: animals treated with a dose of 160 mg/kg (n=12), and E320: animals treated with a dose of 320 mg/kg (n=11), TP: untreated parasitized controls sacrificed on the 60thand^{81st}days (n=20), AT: treated animals (n=44).

V - RELATIVE ORGAN WEIGHTS

V -1- Relative weight of the liver and spleen

V -1-1- Relative weight of the liver

The data summarized in Figure 10 show that, compared to control animals, the relative weight of the liver increases significantly (p<0.001) in untreated parasitized animals. Furthermore, a significant decrease (p<0.01 and p<0.001) in relative liver weight was observed in all treated animals compared to untreated parasitized animals sacrificed on the^{81st}day. This represents a protection rate of 31.86% for the extract.

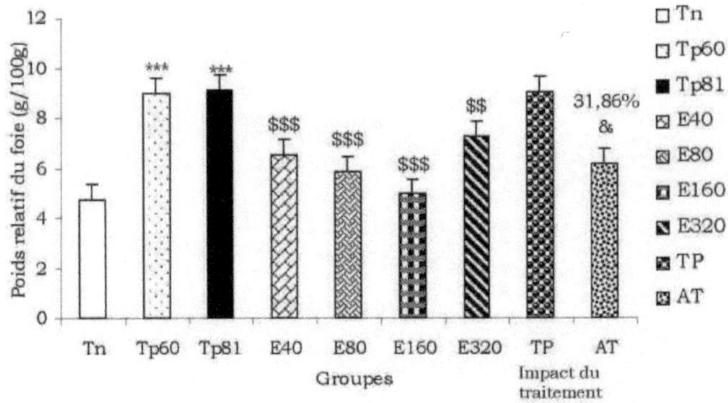

Figure 10: **Effects of the aqueous extract of *Clerodendrum* umbellatum leaves on relative liver weight**

Values represent means ± SEM (standard error of the mean) ***p<0.001 significant difference compared to negative control animals (Tn) $$$ p<0.001 significant difference compared to untreated parasitized animals (Tp81) & p<0.05 significant difference compared to untreated parasitized controls sacrificed on the 60thth and 81st days

Tn: negative control animals (n=10), Tp60: untreated infected animals sacrificed on the 60ththday (n=10), Tp81: untreated parasitized animals sacrificed on the 81stday (n=10), E40: animals treated with a dose of 40 mg/kg (n=10), E80: animals treated with a dose of 80 mg/kg (n=11), E160: animals treated with a dose of 160 mg/kg (n=12), and E320: animals treated with a dose of 320 mg/kg (n=11), TP: untreated parasitized controls sacrificed on the 60thand 81 days (n=20), AT: treated animals (n=44).

V-1-2- Relative weight of the spleen

Regarding the spleen, a significant increase (p<0.05 and p<0.001) in relative spleen weight was observed in untreated parasitized animals compared to control animals. A significant decrease (p<0.05 and p<0.001) was observed in animals treated with the aqueous extract of *Clerodendrum umbellatum* leaves at doses of 40, 80, and 160 mg/kg compared to untreated parasitized animals sacrificed on $^{the\ 81st}$day. In the latter group, a significant increase (p<0.05) was noted compared to untreated parasitized animals sacrificed on the^{60th}day. It appears that the overall protection rate of the extract in treated animals is 25.80%.

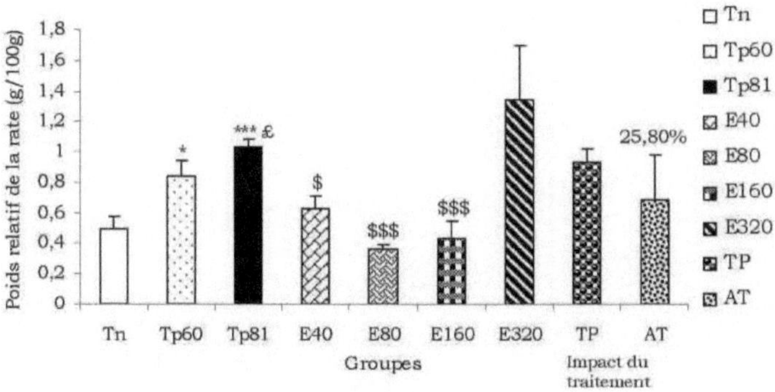

<u>Figure 11:</u> Effects of the aqueous extract of *Clerodendrum umbellatum* leaves on the relative weight of the spleen

Values represent means ± SEM (standard error of the mean)

* $p<0.05$; ***$p<0.001$ significant difference compared to negative control animals (Tn)

£ $p<0.05$: significant difference compared to untreated parasitized animals sacrificed on the^{60th}day \$ $p<0.05$; \$\$\$ $p<0.001$: significant difference compared to untreated parasitized animals (Tp81)

Tn: negative control animals (n=10), Tp60: untreated parasitized animals sacrificed on the 60ththday (n=10), Tp81: untreated parasitized animals sacrificed on the 81stday (n=10), E40: animals treated at a dose of 40 mg/kg (n=10), E80: animals treated with a dose of 80 mg/kg (n=11), E160: animals treated with a dose of 160 mg/kg (n=12), and E320: animals treated with a dose of 320 mg/kg (n=11), TP: untreated parasitized controls sacrificed on the 60thand^{81st}days (n=20), AT: treated animals (n=44).

V-2- Relative weights of the testes, epididymis, and seminal vesicles

V-2-1- Relative weight of the testes

Figure 12 shows the relative weight of the testes. A non-significant increase was observed in untreated, parasitized animals sacrificed on the^{60th}day, and a non-significant decrease in those sacrificed on the^{81st}day, compared to negative control animals. Furthermore, a significant decrease ($p<0.05$) was observed in untreated parasitized animals sacrificed on $^{the\ 81st}$day compared to those sacrificed on the^{60th}day. Additionally, in treated animals, the relative weight of the testes showed a significant increase ($p<0.05$; $p<0.01$) at doses of 40, 80, 160, and 320 mg/kg compared to untreated parasitized animals sacrificed on the^{81st}day. No significant difference was observed between animals treated with the aqueous extract of *Clerodendrum umbellatum* leaves and the negative control animals.

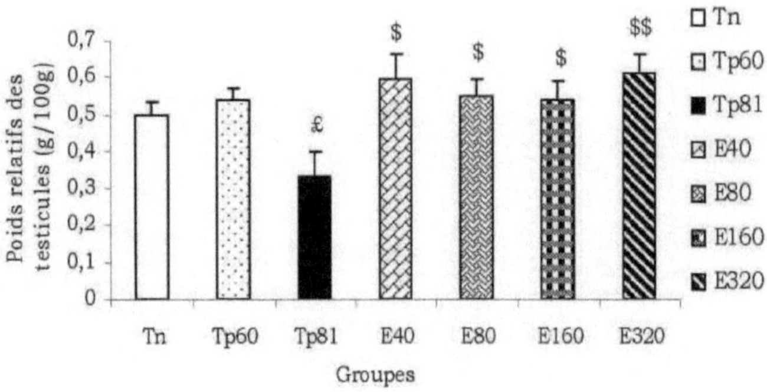

Figure 12: Effects of the aqueous extract of *Clerodendrum umbellatum* leaves on relative testis weight

Values represent means ± SEM (standard error of the mean)

£ p<0.05: significant difference compared to untreated parasitized animals sacrificed on the 60th day $ p<0.05; $$ p<0.01: significant difference compared to untreated parasitized animals sacrificed on the 81st day

Tn: negative control animals (n=10), Tp60: untreated parasitized animals sacrificed on the 60th day (n=10), Tp81: untreated parasitized animals sacrificed on the 81st day (n=10), E40: animals treated with a dose of 40 mg/kg (n=10), E80: animals treated with a dose of 80 mg/kg (n=11), E160: animals treated with a dose of 160 mg/kg (n=12), and E320: animals treated with a dose of 320 mg/kg (n=11).

V-2-2- Relative weight of the epididymis

Figure 13, which shows the effect of the aqueous extract of *Clerodendrum umbellatum* leaves on the relative weight of the epididymis, reveals a significant increase (p<0.05) in the relative weight of the epididymis in untreated parasitized animals sacrificed on the 60th th day and a significant decrease (p<0.001) in untreated parasitized animals sacrificed on the 81st day compared to negative control animals. Comparison among untreated parasitized animals reveals a significant decrease (p<0.05) in the relative weight of the epididymis in untreated parasitized animals sacrificed on the 81th day compared to those sacrificed on the 60th day. In treated mice, a significant increase (p<0.05; p<0.001) was observed at doses of 40, 80, 160, and 320 mg/kg compared to untreated parasitized animals sacrificed on the 81st day. Comparison between animals treated with the aqueous extract of *Clerodendrum umbellatum* leaves and negative control animals revealed a significant increase (p<0.05; p<0.001) at doses of 40, 80, and 320 mg/kg.

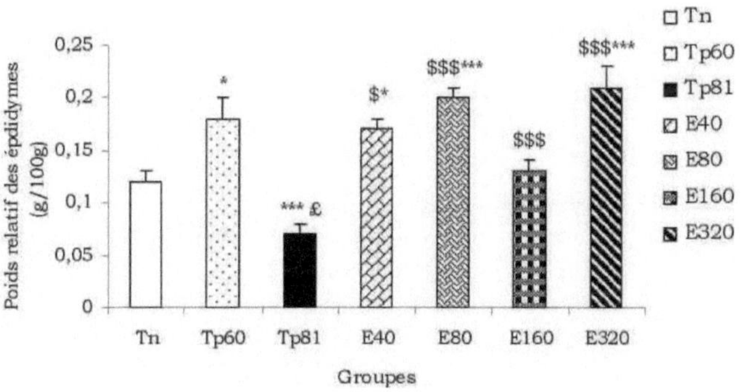

Figure 13: Effects of the aqueous extract of *Clerodendrum umbellatum* leaves on the relative weight of the epididymis

Values represent means ± SEM (standard error of the mean)

* p<0.05; ***p<0.001 significant difference compared to negative control animals

£ p<0.05: significant difference compared to untreated parasitized animals sacrificed on the 60th day

$ p<0.05; $$$ p<0.001 significant difference compared to untreated parasitized animals sacrificed on the 81st day

Tn: negative control animals (n=10), Tp60: untreated parasitized animals sacrificed on day 60 (n=10), Tp81: untreated parasitized animals sacrificed on day 81 (n=10), E40: animals treated with a dose of 40 mg/kg (n=10), E80: animals treated with a dose of 80 mg/kg (n=11), E160: animals treated with a dose of 160 mg/kg (n=12), and E320: animals treated with a dose of 320 mg/kg (n=11).

V-2-3- Relative weight of seminal vesicles

With regard to the relative weight of the seminal vesicles (Figure 14), no significant difference was observed between the negative control group and the groups of untreated parasitized animals. However, a slight increase was noted in untreated parasitized animals sacrificed on the 60th day, and a slight decrease in those sacrificed on the 81st day compared to the negative controls. A non-significant decrease was observed in untreated parasitized animals sacrificed on the 81st day compared to those sacrificed on the 60th day. In treated animals, however, an increase in the relative weight of the seminal vesicles was noted, a significant increase (p<0.05) at the 80 mg/kg dose compared to untreated parasitized animals sacrificed on the 81st day. No significant difference was observed between animals treated with the aqueous extract of *Clerodendrum umbellatum* leaves and negative control animals.

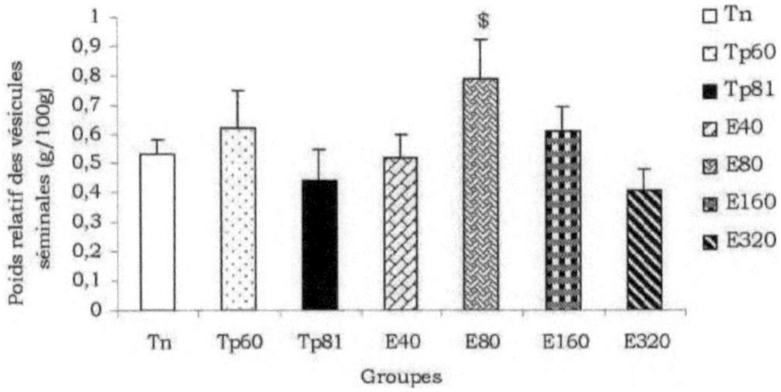

<u>Figure 14</u>: Effects of the aqueous extract of *Clerodendrum umbellatum* leaves on the relative weight of seminal vesicles

Values represent means ± SEM (standard error of the mean)

$ p<0.05: significant difference compared to untreated parasitized animals sacrificed on the[81st]day Tn: negative control animals (n=10), Tp60: untreated parasitized animals sacrificed on the 60th[th]day (n=10), Tp81: untreated parasitized animals sacrificed on the[81st]day (n=10), E40: animals treated at a dose of 40 mg/kg (n=10), E80: animals treated with a dose of 80 mg/kg (n=11), E160: animals treated with a dose of 160 mg/kg (n=12), and E320: animals treated with a dose of 320 mg/kg (n=11).

VI- EPIDIDYMAL ALPHA-GLUCOSIDASE ACTIVITY

Figure 15 shows the effect of the aqueous extract of *Clerodendrum umbellatum* leaves on epididymal alpha-glucosidase activity. The groups of parasitized animals show a decrease in activity that becomes significant (p<0.05) in untreated parasitized animals sacrificed on day 81[e] day compared to the negative control group. There was no significant difference among the groups of parasitized animals. In animals treated with doses of 40, 80, and 160 mg/kg, a significant increase (p<0.05; p<0.001) was observed compared to untreated parasitized animals sacrificed on the 81th day. Comparison between animals treated with the aqueous extract of *Clerodendrum umbellatum* leaves and the negative control animals revealed a significant increase (p<0.05) at the 40 mg/kg dose and a significant decrease (p<0.001) at the 320 mg/kg dose.

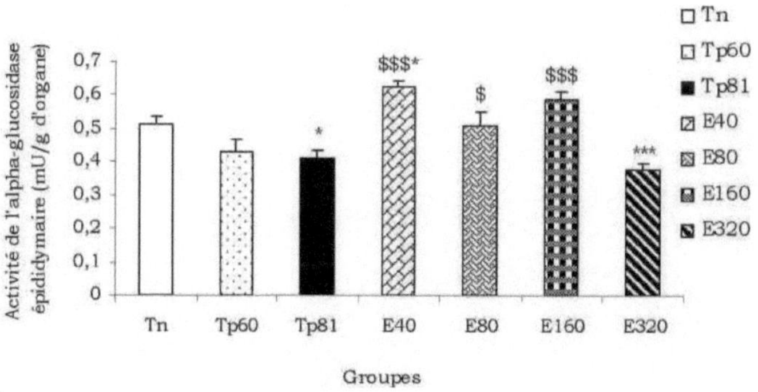

Figure 15: Effects of the aqueous extract of *Clerodendrum umbellatum* leaves on epididymal alpha-glucosidase activity

Values represent means ± SEM (standard error of the mean)

* p<0.05; *** p<0.001 significant difference compared to negative control animals

$ p<0.05; $$$ p<0.001 significant difference compared to untreated parasitized animals sacrificed on the81stday

Tn: negative control animals (n=10), Tp60: untreated parasitized animals sacrificed on day60(n=10), Tp81: untreated parasitized animals sacrificed on day 81 (n=10), E40: animals treated with a dose of 40 mg/kg (n=10), E80: animals treated with a dose of 80 mg/kg (n=11), E160: animals treated with a dose of 160 mg/kg (n=12), and E320: animals treated with a dose of 320 mg/kg (n=11).

VII- SEMEN FRUCTOSE CONCENTRATION

Figure 16 shows that the concentration of seminal fructose increases in untreated parasitized animals; this increase was significant (p<0.05) in those sacrificed on the60thday compared to negative control animals. No notable differences were observed among the untreated parasitized animals. In animals treated with the aqueous extract of *Clerodendrum umbellatum* leaves at a dose of 40 mg/kg, a non-significant increase was noted; whereas at doses of 80, 160, and 320 mg/kg, there was instead a significant decrease (p<0.05; p<0.01) compared to untreated parasitized animals sacrificed on the81stday. No significant difference was observed between animals treated with the aqueous extract of *Clerodendrum umbellatum* leaves and the negative control animals.

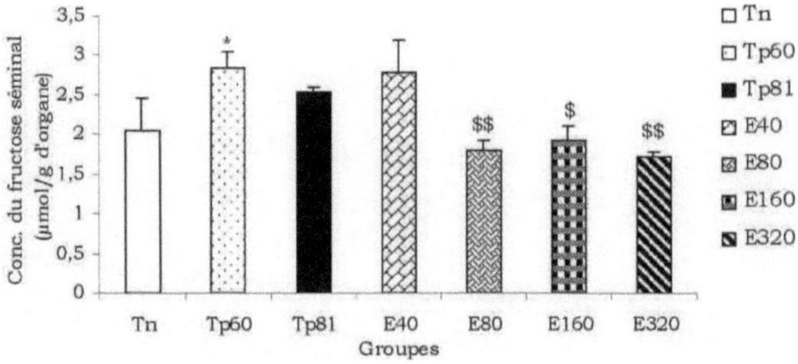

Figure 16: Effects of the aqueous extract of *Clerodendrum umbellatum* leaves on seminal fructose concentration

Values represent means ± SEM (standard error of the mean)

* $p<0.05$: significant difference compared to negative control animals

$ $p<0.05$; $$ $p<0.01$: significant difference compared to untreated parasitized animals sacrificed on the^{81st}day

Tn: negative control animals (n=10), Tp60: untreated parasitized animals sacrificed on the^{60th}day (n=10), Tp81: untreated parasitized animals sacrificed on the^{81st}day (n=10), E40: animals treated with a dose of 40 mg/kg (n=10), E80: animals treated with a dose of 80 mg/kg (n=11), E160: animals treated with a dose of 160 mg/kg (n=12), and E320: animals treated with a dose of 320 mg/kg (n=11).

VIII- SPERM CONCENTRATION

Analysis of the data summarized in Figure 17 revealed a significant increase ($p<0.05$) in sperm count in untreated, parasitized animals sacrificed on the^{60th}day and a non-significant decrease in those sacrificed on $^{the\ 81st}$day compared to negative control animals. A significant decrease ($p<0.01$) was observed in untreated parasitized animals sacrificed on the^{81st}day compared to untreated parasitized animals sacrificed on the^{60th}day. Compared to untreated parasitized animals sacrificed on the 81stday, a significant increase ($p<0.01$) in sperm count was detected in animals treated with the aqueous leaf extract of *Clerodendrum umbellatum* at a dose of 40 mg/kg; whereas at doses of 80 and 160 mg/kg, this increase was not significant. Comparison between animals treated with the aqueous extract of *Clerodendrum umbellatum* leaves and negative control animals revealed a significant increase ($p<0.05$) at the 40 mg/kg dose and a significant decrease ($p<0.05$) at the 320 mg/kg dose.

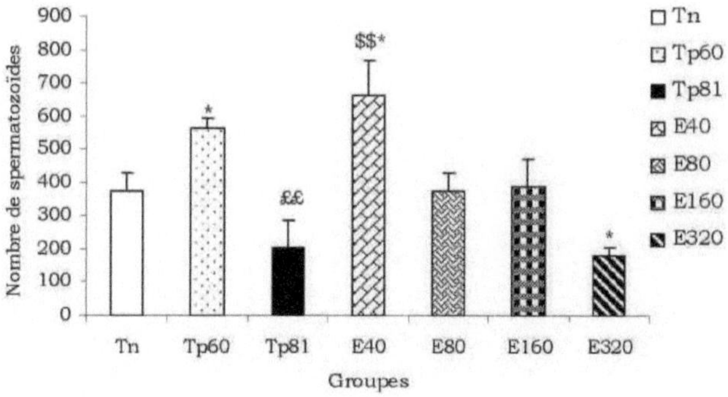

<u>Figure 17</u>: Effects of the aqueous extract of *Clerodendrum umbellatum* leaves on sperm concentration

Values represent means ± SEM (standard error of the mean)

*p<0.05 significant difference compared to negative control animals

££ p<0.01 significant difference compared to untreated parasitized animals sacrificed on the 60thday $$ p<0.05 significant difference compared to untreated parasitized animals sacrificed on the81stday Tn: negative control animals (n=10), Tp60: untreated parasitized animals sacrificed on the 60thday (n=10), Tp81: untreated parasitized animals sacrificed on the81stday (n=10), E40: animals treated at a dose of 40 mg/kg (n=10), E80: animals treated with a dose of 80 mg/kg (n=11), E160: animals treated with a dose of 160 mg/kg (n=12), and E320: animals treated with a dose of 320 mg/kg (n=11).

IX- HISTOLOGICAL ANALYSIS OF THE LIVER

Histological sections of the livers of untreated, parasitized animals were compared with those of negative control animals. Meanwhile, histological sections from animals treated with an aqueous extract of *Clerodendrum umbellatum* leaves were compared with those of untreated, parasitized animals and with those of negative control animals.

Figure 1 shows the liver of a control mouse at 100x magnification. The hepatic parenchyma has a normal architecture with well-differentiated centrilobular veins and portal spaces. Hepatocytes are visible.

Figure 2 shows the liver of an untreated parasitized mouse at 100x magnification. The hepatic parenchyma is dotted with numerous bilharzian granulomas. At 400× magnification (Figure 3), the granuloma is clearly visible with concentric layers of inflammatory cells and collagen fibers: this is the fibrotic granuloma.

Figure 4 shows the liver of a mouse treated with an aqueous extract of *Clerodendrum umbellatum* leaves at a dose of 40 mg/kg at 100× magnification. The parenchyma shows a few bilharzian granulomas that have not completely disappeared after treatment. At a dose of

160 mg/kg (plate 5), there are no longer any clearly distinct granulomas; only traces of collagen fibers remain: granuloma involution is observed.

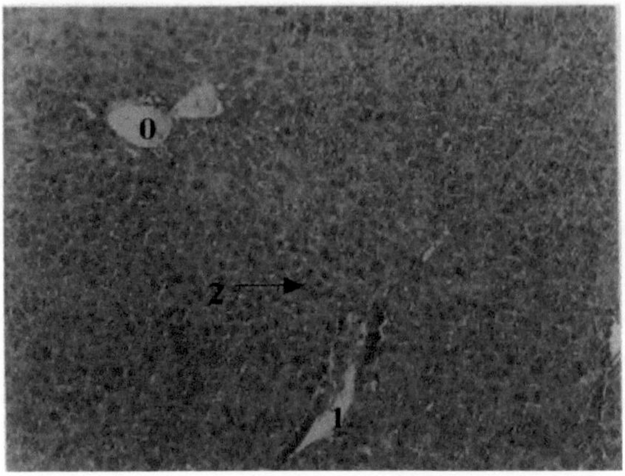

Figure 1: Liver of a control mouse (HE x 100)

Liver parenchyma with normal architecture, showing distinct centrilobular veins (0), portal spaces (1), and hepatocytes (2).

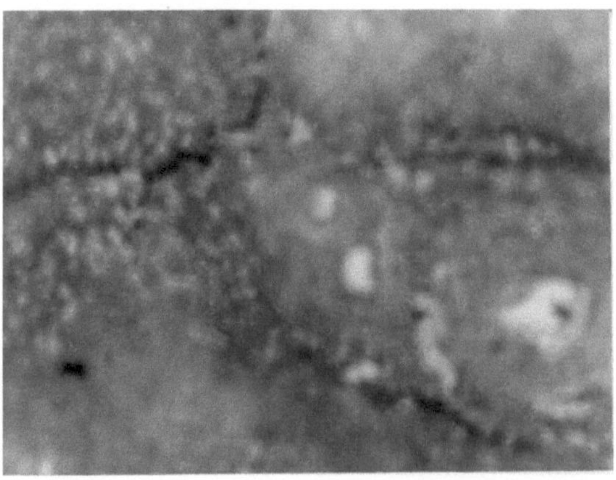

Figure 2: Liver of an untreated parasitized mouse (TM x 100) Hepatic parenchyma dotted with numerous granulomas (3).

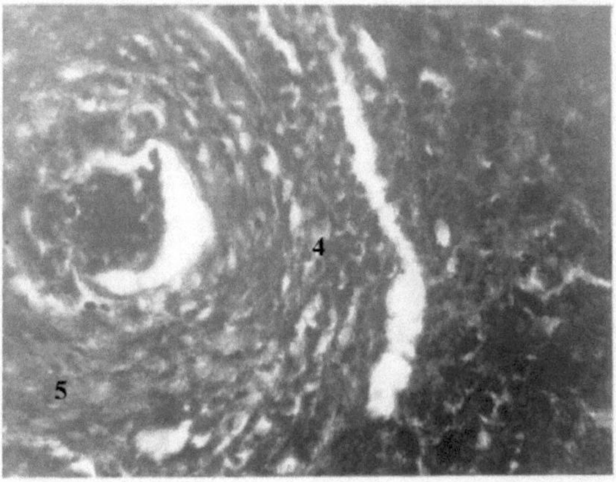

Figure 3: Liver of an untreated parasitized mouse (TM x 400)

Concentric layers of inflammatory cells (4) and collagen fibers (5).

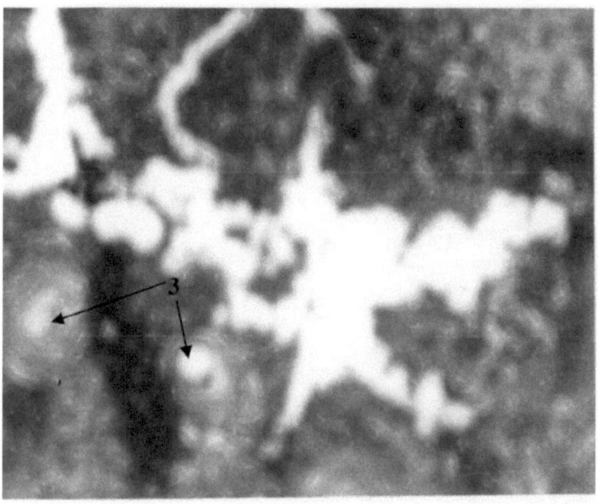

Plate 4: Liver of a mouse treated with an aqueous extract of *C. umbellatum* leaves at a dose of 40 mg/kg (TM x 100).

Fewer granulomas (3) located at the periphery.

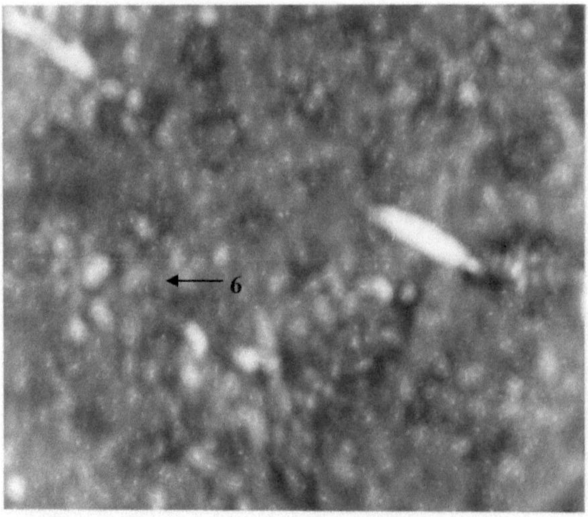

<u>Figure 5</u>: Liver of a mouse treated with an aqueous extract of *C. umbellatum* leaves at a dose of 160 mg/kg (TM × 100).

No granulomas; presence of collagen fibers (6).

Discussion

Infection of male mice with cercariae from *Biomphalaria pfeifferi* snails resulted in the shedding of *S. mansoni* eggs in the feces and their presence in the liver, observed on the 60th day post-infection. It is known that in schistosomiasis, pathogenicity is linked to the presence of parasite eggs in the definitive host's body (Golvan, 1983). During their trans-tissue migration, these eggs release toxic substances and free radicals that are responsible for the formation of schistosomal granulomas or schistosomiasis granulomas (Grossetete et al., 1989), as observed in our histological sections. Indeed, the eggs present in the intrahepatic branches of the portal system cause necrosis of the vascular wall and migrate into the portal spaces, where they are surrounded by a reaction consisting of histiocytes, eosinophils, and subsequently a proliferation of connective tissue (Gentillini et al., 1982). Infection in mice was also manifested by the presence of schistosomes in the circulatory system. Like the eggs, these schistosomes release toxic substances and free radicals resulting from their metabolism, which are responsible for all the clinical manifestations of the initial phases of schistosomiasis (Houin, 1981).

When treating infected mice with an aqueous extract of *Clerodendrum umbellatum* leaves, we observed a mortality rate of 8.33%, compared to 56.52% in untreated infected mice. Furthermore, treatment of the mice with the aqueous extract of *Clerodendrum umbellatum* leaves resulted in an overall reduction in the number of eggs in the feces by 54.48% and in the liver by 22.74% in all treated mice compared to the parasitized mice. A reduction in the number of schistosomes of 88.45% was also observed in these animals. Hamed and Hetta (2005) had obtained, in their study on the efficacy of the ethanolic extract of *Citrus reticulata* roots and Mirazid in the treatment of *Schistosoma* mansoni schistosomiasis*, mansoni*, a 63.66% reduction in the number of schistosomes for *Citrus reticulata* (administered orally for three consecutive weeks at a dose of 10 µg/ml) and 81.10% for Mirazid (administered orally for three consecutive days at a dose of 600 mg/kg). Koko et al. (2005), by treating mice infected with the Sudanese strain of *Schistosoma mansoni* with a single dose of 200 mg/kg, found a worm reduction rate of 73% for *Balanites aegyptiaca* and 42.7% for praziquantel, respectively. The 2004 study by Botros et al. on the demonstration of myrrh's anti-schistosomal activity in animal experiments revealed that, following treatment of mice and hamsters infected with the Egyptian, Puerto Rican, and Brazilian strains of *Schistosoma mansoni*, a reduction in worms ranging from 19% to 36% with myrrh and Mirazid (administered orally for 5 days at doses ranging from 180 to 10,000 mg/kg) and 94% with praziquantel (200 mg/kg for 5 days). Compared to the worm reduction rates obtained in the studies listed above, it is evident that the rate for *Clerodendrum umbellatum*, which is 88.45%, is significantly higher than the others and very close to the reduction rate of praziquantel (reference product) obtained by Botros et al. in 2004. These results suggest that the aqueous leaf extract of *Clerodendrum umbellatum* possesses an anti-schistosomal effect, which manifested in this study as a decrease in the number of eggs and schistosomes following treatment. This reduction can be explained by the presence in the extract of alkaloids, known to have anti-inflammatory (Venkatesh et al., 1999) and antiparasitic properties (Lhoste et al., 1993; Pamo et al., 2003); triterpenes, which have anti-inflammatory properties (Venkatesh et al., 1999); and flavonoids and tannins, which are known to possess anti-inflammatory (Venkatesh et al., 1999), antioxidant (Hennebelle et al., 2004; Hara et al., 2004; Ojewole, 2004), and antiparasitic properties (Lhoste et al., 1993; Pamo et al., 2003). The presence of these compounds in the aqueous extract of *Clerodendrum umbellatum* leaves was demonstrated during the phytochemical testing conducted in this study. The extract's anti-inflammatory property would limit the inflammatory reaction caused by the presence of eggs, which leads to the formation of schistosomes. The antioxidant property of the compounds would limit the action of toxic substances and free radicals released by schistosomes and their eggs (Hamed and Hetta, 2005); which would allow the liver to perform its role in detoxifying these endogenous and exogenous toxic substances, as the presence of these foreign substances could impair the liver's detoxification capacity (Sheweita et al., 1998). The antiparasitic property could be explained by the significant reduction in the number of schistosomes observed after treatment of infected mice. Furthermore, since schistosomes are the source of the eggs, the reduction in the number of schistosomes would explain the reduction in eggs following treatment.

During our experiment, we observed that the body weight of the animals varied only slightly during the first six weeks but began to decline starting in the seventh week (^{49th}day). This decline was statistically significant ($p<0.05$) by the eighth week (^{60th}day). These results are consistent with the findings of Fiore et al. (1996), who reported that *Schistosoma mansoni* infection leads to a reduction in body weight eight weeks after infection. In *S. mansoni*, the prepatent period (the time between cercaria penetration and egg release) is 50 to 55 days (Malek, 1980). The decrease in body weight during this period corresponds to the onset of egg release by female parasites. This weight loss could also result from anemia caused by the degradation of hemoglobin in the host's red blood cells by schistosomes (Wasilewski et al., 1996) on the one hand, and on the other hand by the feeding habits of these worms, which consume the host's glucose for their metabolism (Skelly et al., 1998). This glucose consumption would create a state of hypoglycemia that could explain the decrease in body weight. Infection in the mice was also manifested during this same period by a significant increase ($p<0.05$; $p<0.001$) in the relative weight of the liver (hepatomegaly) and spleen (splenomegaly) in the infected mice compared to the control mice. Ahmed and Gad in 1995 observed a correlation between decreased body weight and egg shedding as well as hepatomegaly following *S. mansoni* infection in mice, but six weeks post-infection. The hepatomegaly observed in these infested mice is likely the result of the presence of schistosomal granulomas in the liver. Splenomegaly, on the other hand, is thought to result from passive venous congestion and hyperplasia of the spleen's reticuloendothelial system (Chen and Mott, 1988). It may also be due to splenic hyperfunction resulting from the excessive elimination of red blood cells destroyed by schistosomes and/or the destruction of schistosome eggs or adult worms that may have been present there (Kuntz et al., 1978; 1981).

Treatment with an aqueous extract of *Clerodendrum umbellatum* leaves resulted in a slight increase in body weight in animals treated at a dose of 160 mg/kg, a significant decrease ($p<0.001$) in relative liver weight in all treated animals, representing an overall protection rate of 31.86% for the extract-treated animals compared to the parasitized animals. A significant decrease ($p<0.05$; $p<0.001$) in relative spleen weight was observed in animals treated with doses of 40, 80, and 160 mg/kg. At the 320 mg/kg dose, splenomegaly persisted, likely due to the high number of eggs remaining in the feces and liver of these animals after treatment. Thus, the percentage of protection provided by the extract regarding the spleen is 25.80%. These improvements in pathologies (hepatomegaly and splenomegaly) may be explained by the reduction in the number of eggs and schistosomes observed after treating infested mice with the aqueous extract of *Clerodendrum umbellatum* leaves, a consequence of the presence in the extract of compounds with anti-inflammatory, antioxidant, and antiparasitic properties.

With regard to reproduction, a non-significant increase in the relative weight of the testes, seminal vesicles, and epididymis was observed in untreated, infested mice sacrificed on the^{60th}day after infestation, compared to the controls. In untreated infected mice sacrificed on the^{81st}day after infestation, a significant decrease ($p<0.05$) in the relative weight of these same organs was observed compared to control mice. Since the development and secretion of these organs are androgen-dependent, these results could be explained on the ^{60th}day by the unchanged testosterone level, as the disease had not yet affected reproductive function at the very onset of its establishment. Jatsa (1999) had reported on $^{the\ 60th}$ post-infestation, testosterone levels comparable to those of control mice in mice with *S. intercalatum* schistosomiasis. The decrease in the relative weight of these reproductive organs on the^{81st}day post-infestation could be explained by a decrease in testosterone levels, as the progression of the disease over time would affect testosterone levels. Indeed, Marzouki and Amin reported in 1997 that *Schistosoma mansoni* schistosomiasis in mice leads to a decrease in testosterone levels by the^{14th}week post-infestation. Furthermore, this disease may affect FSH and LH levels and subsequently lead to a decrease in testosterone (Lansaud-Soukate et al., 1991). The significant decrease ($p<0.05$) in epididymal alpha-glucosidase activity, as well as the significant decrease ($p<0.01$) in sperm count on the^{81st}day, are consistent with these results. Furthermore, the decrease in alpha-glucosidase activity would explain the reduction in sperm count. Seminal fructose concentration, however, remains elevated in parasitized animals compared to healthy control animals. This result, which does not align with the decrease in the relative weight of the seminal vesicles on the^{81st}day, cannot be explained by the decrease in testosterone levels. Jatsa (1999) had noted that there is a negative correlation between the weight of the seminal vesicles and fructose concentration in parasitized mice. Indeed, in rats and mice, fructose is primarily synthesized by the "coagulating glands" attached to the seminal vesicles and by the dorsolateral surface of the prostate; whereas the seminal vesicles primarily synthesize citric acid. These coagulating glands contain the enzymatic machinery required (aldose reductase and sorbitol dehydrogenase) to convert glucose from the bloodstream into fructose (Samuels and Harding, 1962). , fructose concentration remained high in parasitized mice, even though it is known that schistosomes use glucose drawn from the host's bloodstream for their metabolism (Skelly et al., 1998). This consumption of blood glucose should normally lead to a decrease in its concentration and consequently a decrease in the fructose level. But this is not what was observed. Furthermore, in 1995, Ahmed and Gad noted, by measuring certain metabolic enzymes in the livers of mice infested with *S. mansoni*, an increase in the activity of phosphofructokinase and pyruvate kinase (glycolysis enzymes) and a reduction in the activity of citrate synthase (a Krebs cycle enzyme). Subsequently, Hamed and Hetta (2005) reported a reduction in succinate dehydrogenase (an enzyme of the Krebs cycle). Thus, during an infestation with

In *S. mansoni*, glycolysis is stimulated, leading to a decrease in glucose levels and the inhibition of the Krebs cycle in favor of fermentation. This is due to the state of hypoxia created by toxic substances and free radicals released by the schistosomes and/or their eggs. Under normal conditions, when blood glucose levels rise following the consumption of carbohydrate-rich foods, this glucose is either used by the liver for glycogen synthesis or broken down into pyruvate via glycolysis. Under aerobic conditions, pyruvate enters the Krebs cycle to provide metabolic energy in the form of ATP (adenosine triphosphate); under anaerobic conditions, it is converted into lactate (as in prolonged physical exertion). Hamed and Hetta (2005) noted that *S. mansoni* infection in mice also leads to increased activity of glucose-6-phosphatase, a glycogenolysis enzyme that catalyzes the conversion of glucose-6-phosphate into glucose. This occurs following an increase in calcium concentration in the cytoplasm of cells. The glucose thus released could either enter glycolysis to be broken down or follow the polyol pathway to be converted into fructose in the "coagulating glands." This would explain the increased fructose levels in parasitized mice.

Regarding the effect of treatment with an aqueous extract of *Clerodendrum umbellatum* leaves on reproductive function parameters in parasitized mice, a significant increase in the relative weight of the testes and epididymis was observed in treated animals at all doses, a significant increase at the 80 mg/kg dose and a non-significant increase at the 40 and 160 mg/kg doses in the relative weight of the seminal vesicles, a significant increase in alpha-glucosidase activity at the 40, 80, and 160 mg/kg doses, a significant increase in sperm count at the 40 mg/kg dose and a non- ly significant increase at the 80 and 160 mg/kg doses, a significant decrease in fructose concentration at the 80, 160, and 320 mg/kg doses compared to untreated, parasitized animals sacrificed on the^{81st}day. When these various results are compared to those of healthy controls, no statistically significant difference emerges overall. This indicates a trend toward normalization of reproductive function parameters, thus demonstrating an improvement in the damage caused by the infection. This improvement may be explained by the normalization of testosterone levels following the reduction in the number of eggs and worms observed after treatment, thanks to the action of bioactive compounds with anti-inflammatory, antioxidant, and antiparasitic activity present in the extract.

Histologically, liver sections from parasitized mice show parenchyma dotted with numerous granulomas of varying sizes. Granulomas are characteristic of schistosomiasis and result from the embolization of schistosome eggs in the hepatic venous system. They constitute distinct chronic forms of inflammation and are characterized by an aggregation of macrophages around the eggs (Amiri et al., 1992), as observed in our histological sections.

In mice treated with a dose of 160 mg/kg, histological sections no longer showed discrete granulomas but rather traces of collagen fibers. This finding is consistent with the marked reduction in the number of eggs in the feces and liver, the absence of schistosomes, and the decrease in relative liver weight observed in these animals following treatment. The disappearance of the granulomas could be explained by the anti-inflammatory action of the alkaloids, flavonoids, tannins, and triterpenes present in the extract. Indeed, this 160 mg/kg dose appears to have been more effective than the others because the initial egg load in the feces was the lowest. This suggests that the extract may be able to cure less severe *S. mansoni* schistosomiasis at the outset.

CONCLUSION AND OUTLOOK

At the conclusion of this study, it appears that hepatomegaly, splenomegaly, and reproductive dysfunction caused by *S. mansoni* schistosomiasis in male mice improved following treatment with the aqueous extract of *Clerodendrum umbellatum* leaves. This improvement is supported by a significant reduction in the number of eggs (54.48% in feces and 22.74% in the liver) and schistosomes (88.45%). This is due to the presence in the extract of compounds with anti-inflammatory, antioxidant, and antiparasitic activities.

Thus, the anti-helminthic effect advocated by traditional practitioners has been verified for *Schistosoma mansoni*, with a dose of 80 mg/kg found to be effective.

Looking ahead, we plan to:

- the treatment of *Schistosoma mansoni* schistosomiasis with praziquantel, considered the standard drug, in order to compare its results with those obtained in this study;
- measuring liver function markers;
- a fertility study;
- the study of the preventive effect and toxicity of the extract
- and to test the efficacy of methanol and methylene chloride/methanol extracts against *Schistosoma mansoni* schistosomiasis, as well as their fractions.

References

1. **Abdalla K.F., Abdel-Aziz S.M., el-Fakahany A.F., el-Hamshary A.S., Afifi L.M.**, (1994). Effect of praziquantel on sex hormone levels in murine schistosomiasis *mansoni. Journal of the Egyptian Society of Parasitology,* **24** (3): 627-632.

2. **Adjanohoun J.E., Aboubakar N., Dramane K., Ebat M.E. et al.**, (1996). Traditional medicine and pharmacopoeia: contribution to ethnobotanical and floristic studies in Cameroon. CNPMS Yaoundé, 596 p.

3. **Ahmed S.A., Gad M.Z.**, (1995). Effect of schistosomal infection and its treatment on some key enzymes of glucose metabolism in mouse livers. *Arzneimittelforschung*, **45** (12): 1324-8.

4. **Amano T., Freeman G.L., Colley D.G.**, (1990). Reduced reproductive efficiency in mice with Schistosoma *mansoni* and in uninfected pregnant mice injected with antibodies against *Schistosoma mansoni* soluble egg antigens. *American Journal of Tropical Medicine and Hygiene*, **43** (2): 180085.

5. **Amiri P., Locksley R.M., Parslow T.G., Sadick M., Rector E., Ritter D., McKerrow J.H.**, (1992). Tumor necrosis factor alpha restores granulomas and induces parasitic egg-laying in schistosome-infected SCID mice. *Nature*, **356**: 604-607.

6. **Aquila S., Gentile M., Middea E., Catalano S., Ando S.**, (2004). Autocrine Regulation of Insulin Secretion in Human Ejaculated Spermatozoa. *Endocrinology* **146** (2): 552-557.

7. **Brooker S., Donnelly C.A., Guyatt H.L.**, (2000). Estimating the number of helminthic infections in the Republic of Cameroon from data on infection prevalence in school children. *Bulletin of* the WHO, **79**: 1456-1465.

8. **Botros S., William S., Ebeid F., Cioli D., Katz N., Day T.A., Bennett J.L.**, (2004). Lack of evidence for an antichistosomal activity of Myrrh in experimental animals. The *American Journal of Tropical Medicine and Hygiene*, **71** (2): 206-210.

9. **Botros S., Sayed H., El-Dusoki H., Sabry H., Rabie I., El-Ghannam M., Hassanein M., El-Wahab Y.A., Engels D.**, (2005). Efficacy of Mirazid in comparison with praziquantel in Egyptian *Schistosoma* mansoni-infected schoolchildren and households. The *American Journal of Tropical Medicine and Hygiene*, **72** (2): 119-123.

10. **Bourée, P.** (1989). Department of Pharmacology—Tropical Diseases. Biâtre Hospital and Paris-Sud Faculty of Medicine. 2nd ed., Paris, 279 pp.

11. **Bourel M.**, (1991). Hepatology, Ed. Bordas, Ellipses, Paris, 383 pp.

12. **Capron A., Capron M.**, (1978). Immunity in Parasitic Diseases. *Inserm,* **79**: 105–114.

13. **Capron M., Capron A., Torpier G., Bazin H., Bout D., Joseph M.**, (1978). Eosinophil-dependent cytotoxicity in rat schistosomiasis. Involvement of IgG 2a antibody and role of mast cells. *European Journal of Immunology*, **8** (2): 127-133.

14. **Carreau S., Foucault P., Drosdowsky M.A.**, (1994). The Sertoli cell; comparative functional aspects in rats, pigs, and humans. *Annales d'endocrinologie*. **55**: 203-220.

15. **Carreau S.**, (1996). Paracrine control of human Leydig cell and Sertoli cell functions. *Folia Histochemica et Cytologica*. **34** (3): 111-119.

16. **Cavalière H., Leile Z., Medeiros-Neto G.**, (1986). Serum immunoreactive somatomedin-C levels in growth failure and delayed puberty associated with chronic hepatosplenic schistosomiasis. *Clinical Endocrinology*, **24**: 617-626.

17. **Chae S., Kim J.S., Kang K.A., Bu H.D., Lee Y., Hyun J.W., Kang S.S.**, (2004). Antioxidant activity of Jionoside D from *Clerodendrum trichotomum*. *Biology and Pharmacology Bulletin*, **27** (10): 1504-1508.

18. **Chae S., Kim J.S., Kang K.A., Bu H.D., Lee Y., Seo Y.R., Hyun J.W., Kang S.S.**, (2005). Antioxidant activity of isoacteoside from *Clerodendrum trichotomum*. *Journal of Toxicology and Environmental Health*, **68** (5): 389-400.

19. **Chatelain P.G., Sanchez P., Saez J.M.**, (1991). Growth hormone and insulin-like growth factor-1 increase testicular luteinizing hormone receptors and steroidogenic responsiveness in growth hormone-deficient dwarf mice. *Endocrinology*, **123**: 1857-1862.

20. **Cheek M., Pollard B.J., Darbyshire I., Onana J.M., Wild C.**, (2004). The Plants of Kupe, Mwanenguba, and the Bakossi Mountains, Cameroon. Published by the Royal Botanic Gardens, Kew, and the National Herbarium of Cameroon, 508 pp.

21. **Cheever A.W.**, (1968). Conditions affecting the accuracy of potassium hydroxide- -digestion techniques for counting *Schistosoma mansoni* eggs in tissues. *Bulletin of the World Health Organization*, **39**: 328-331.

22. **Chen M.G., Mott K.E.**, (1988). Progress in the assessment of morbidity due to *Schistosoma mansoni* infection. A review of recent literature. Tropical Diseases Bulletin. Vol. 35, No. 10, 56 pp.

23. **Combes C., Fournier A., Mingyl X.**, (1987). Schistosomes. *Pour la Science*, **116**: 80-88.

24. **Dadoune J.P., Hadjisky P., Siffroi J.P., Vendrely E.**, (1990). From Biology to Clinical Practice: Histology, Ed. Médecine Sciences Flammarion, Paris, pp. 278–287.

25. **Dadoune J.P., Demoulin A.**, (1991). Structure and Function of the Testis. In: Reproduction in Mammals and Humans. **Thibault C, Levasseur M.C.** (Eds). Edition Marketing, Paris, pp. 221–250.

26. **Di Fiore M.S.H.**, (1965). Atlas of Human Histology, 2nd ed. (Lea & Febiger), Philadelphia, 224 pp.

27. **Drosdowsky M., Mahoudeau J.A.**, (1980). Physiology and investigation of the endocrine function of the testis. Encyclopedia of Medicine and Surgery; Paris; Glands 10032, pp. 1–10.

28. **Duvall R.H., Dewitt W.B.**, (1967). An improved perfusion technique for recovering adult schistosomes from laboratory animals. *American Journal of Parasitology*, **7**: 293-297.

29. **Dyn M., Fawcett D.W.**, (1970). The blood-testis barrier in the physiological compartmentation of the seminiferous epithelium. *Biol. Reprod.*, **3**: 308-326.

30. **El-Ridi A.M.S., Aboul-Magd L.A., Edris A., Rashed S.M., Hamadto H.A., Abdalla K.F.**, (1987). Variation in serum testosterone levels in male patients with schistosomiasis. *Journal of the Egyptian Society of Parasitology*, **17**: 87-92.

31. **Evengard B.,** (1989). Schistosomiasis: Immunological, serological, and clinical aspects. *Scandinavian Journal of Infectious Diseases, Supplement* 63, 72 pp.

32. **Flisser A., Mclaren D.J.**, (1989). Effect of praziquantel treatment on lung-stage larvae of *Schistosoma mansoni in vivo. Parasitology,* **98**: 203-211.

33. **Fiore M., Mozoni R., Alleva E., Aloe L.**, (1996). *Schistosoma mansoni*: Influence of infection on mouse behavior. *Experimental Parasitology* **83**: 4654.

34. **Fournier-Delpech S., Thibault C.**, (1991). Acquisition of sperm fertilizing capacity:

Epididymal maturation, accessory glands, and capacitation. In: Reproduction in Mammals and Humans. **Thibault C, Levasseur M.C.** (Eds). Edition Marketing, Paris, pp. 251–271.

35. **Gelfand M., Ross C.M.D., Castle W.M., Weber M.C.** (1970). Schistosomiasis of the male pelvic organs. Severity of infection as determined by tissue digestion and histological methods in 300 cadavers. *American Journal of Tropical Medicine and Hygiene*, **19** (5): 779-784.

36. **Gentillini M., Duflo Danis M., La Gardere B., Richard Lenoble D.C.** (1982). Schistosomiasis. Tropical Medicine, Flammarion Medicine Science. Paris. pp. 202–207.

37. **Girgis S.M., Wassef N.F.**, (1980). Schistosomiasis and azoospermia. *Archives of Andrology*, **5** (4): 369-372.

38. **Golvan Y.J.**, (1983). Elements of Medical Parasitology, 4th ed., Flammarion, Paris, p. 571.

39. **Grizard G., Jimenez C.**, (1997). Semen analysis in the evaluation of male fertility. Department of Developmental and Reproductive Biology, University Hospital Center, Clermont-Ferrand, France, **7**: 496-504.

40. **Grossetete G., Diabate I., Pichard E., Ketta S., Vignon D., Meyran M.**, (1989). Cutaneous manifestations of schistosomiasis: A report on 24 cases in Mali. *Bulletin of the Society of Exotic Pathology*, Affiliates, **82**: 225–232.

41. **Grootegoed J.A., Den Boer P.J., Mackenbach P.**, (1989). Role of Sertoli cells in spermatogenesis. In: Perspectives in Andrology, Sereno Symposia, Series M (Ed.). *Rave Press, New York*, **53**: 167–173.

42. **Guyton A.C., Hall J.E.**, (1996). Textbooks of Medical Physiology, 9th ed., Saunders, USA, pp. 315–982.

43. **Hamed Manal A., Hetta Mona H.**, (2005). Efficacy of *Citrus reticulata* and Mirazid in the treatment of *Schistosoma mansoni*. Memórias do Instituto Oswaldo Cruz, **100** (7): 771–778.

44. **Hanukoglu I.**, (1992). Steroidogenic enzymes: structure, function, and role in the regulation of steroid hormone biosynthesis. *Journal of Steroid Biochemistry and Molecular Biology*, **43**: 779–804.

45. **Hara M., Fujinagas M., Kuboi T.**, (2004). Radical scavenging activity and oxidative

modification of *Citrus dehydrin. Plant Physiology and Biochemistry*, **42**: 657–662.

46. **Hazekamp A., Verpoorte R., Panthong A.**, (2001). Isolation of a bronchodilator flavonoid from the Thai medicinal plant *Clerodendrum petasites. Journal of* Ethnopharmacology, **78** (1): 45-9.

47. **Hennebelle T., Sahpaz S., Bailleul F.**, (2004). Plant polyphenols: sources, uses, and potential in combating oxidative stress. *Phytotherapy*, **1**: 1-5.

48. **Honore L.H., Coleman G.U.**, (1975). Solitary epididymal schistosomiasis. *Canadian Journal of Surgery*, **18** (5): 479-483.

49. **Houin R.**, (1981). Schistosomiasis. Encyclopédie-Médico-Chirurgicale, Paris. Infectious Diseases, 8111A10-5.

50. **Huhtaniemi I., Toppari J.**, (1996). Hormonal regulation of spermatogenesis. In: Advances in Andrology: Male Endocrinology;

Drosdowsky M.A., Belaisch J., Vermeulen A., (Eds). Douin Venily (France), 63–74.

51. **Hutchinson J., Dalziel J.M.,** (1963). Flora of West Tropical Africa, Vol. II, The Whitefriars Press, Ltd., London and Tonbridge. 544 pp.

52. **Ihekwaba F.N.**, (1992). Schistosomiasis of the testis. *Central African Journal of Medicine*, **38** (3): 123-127.

53. **Jatsa B. H.**, (1999). Schistosomiasis caused by *Schistosoma intercalatum* Fisher, 1934 and its impact on certain testicular and ovarian functions in the mouse *Mus musculus* L. Doctoral dissertation, University of Yaoundé I, 168 pp.

54. **Johnson M.H., Everitt B.J.**, (1988). Essential Reproduction. Blackwell Scientific Publications, London (3rd ed.), 377 p.

55. **Jost A.**, (1991). Sexual differentiation. In: Reproduction in Mammals and Humans; **Thibault C., Levasseur M.C**. (Eds). Edition Marketing, Paris, 189–203.

56. **Kabatereine N.B., Kemijumbi J., Ouma J.H., Sturrock R.F., Butterworth A.E., Madsen H., Ornbierg N., Dunne D.W., Vennervald B.J.**, (2003). Efficacy and side effects of praziquantel treatment in a highly endemic *Schistosoma mansoni* focus at Lake Albert, Uganda. *Transactions of the Royal Society of Tropical Medicine and Hygiene,* **97**: 599–603.

57. **Katz N., Coelho P.M.Z., Pelligrino J.,** (1970). Evaluation of Kato's quantitative

method through the recovery of *S. mansoni* eggs. *Journal of Parasitology*, **56**: 1032-1033.

58. **Koko W.S., Abdalla H.S., Galal M., Khalid H.S.**, (2005). Evaluation of oral therapy for Mansonial schistosomiasis using a single dose of *Balanites aegyptiaca* fruits and praziquantel. *Fitoterapia*, **76**: 30-34.

59. **Koueta H., Kamtchouing P.**, (1992). Sites of testicular estrogen biosynthesis. *Camerounian Journal of Biological and Chemical* Sciences, **2**: 131–150.

60. **Kuntz R.E., McCullough B., Moore J.A., Huang T.C.**, (1978). Experimental infection with *Schistosoma intercalatum* (Fisher, 1934) in the chimpanzee (*Pan troglodytes*) and the gibbon (*Hylobates lar*). *American Journal of Tropical Medicine and Hygiene,* **27**: 632–634.

61. **Kuntz R.E., Moore J.A., Huang T.C.**, (1981). Variable *Schistosoma intercalatum* infection in cynomolgus macaques (*Macacca fascicularis*). *Journal of Medical Primatology,* **10**: 175-179.

62. **Lansoud-Soukate J., Leonardelli J., Torpier G., Croix D., Capron A.**, (1991). Role of *Schistosoma mansoni* bilharziasis in male hypogonadism. *Pathol Biol* (Paris). **39** (7): 981-5.

63. **Lejeune H., Avallet O., Saez J.M.**, (1994). Functional interactions between somatic cells of the male gonad. *Contraception, Fertility, and Sexuality*, **22** (9): 539-547.

64. **Lhoste P., Dolle V., Rousseau J., Soltner D.**, (1993). Manual of Animal Husbandry in Warm Regions. Livestock Systems. Livestock Handbook Series. Ministry of Cooperation, Paris, 288 pp.

65. **Ma T.K.F., Srigley J.R.**, (1995). Adenocarcinoma of the prostate and schistosomiasis: a rare association. *Histopathology*, **27**: 187–189.

66. **Malek E.A.**, (1980). Snail-Transmitted Parasitic Diseases. CRC Press. 307 pp.

67. **Marieb E.N.**, (1999). Human Anatomy and Physiology. (Eds) De Boeck University, Paris, Brussels (4th ed.), 1194 pp.

68. **Marzouki Z.M., Amin A.M.**, (1997). The effect of *Schistosoma mansoni* infection on testicular lipids in mice. *Journal of the Egyptian Society of* Parasitology, **27** (2): 581-95.

69. **Mercier-Bodard C., Alfsen A., Baulieu E.E.**, (1970). Sex steroid binding plasma protein (SBP). Karolinska symposia on research methods in reproductive endocrinology

2nd symposium, steroid assay by protein binding, 204–224.

70. **Miller R.E., McConville M.J., Woodrow I.E.**, (2006). Cyanogenic glycosides from the rare Australian endemic rainforest tree *Clerodendrum grayi* (Lamiaceae). *Phytochemistry*, **67** (1): 43-51.

71. **Nan H., Wu J., Zhang S.**, (2005). A new phenylethanoid glycoside from *Clerodendrum inerme*. *Pharmazie*, **60** (10): 798-9.

72. **Nash T.E., Cheever A.W., Ottensen E.A., Cook J.A.**, (1982). Schistosome infections in humans: Perspectives and recent findings. *Annals of Internal Medicine*, **97**: 740-754.

73. **Ojowele J.A.**, (2004). Potentiation of the anti-inflammatory effect of *Anacardium occidentale* (Linn.) stem-bark aqueous extract by grapefruit. *Methods and Findings in Clinical* Pharmacology, **26**: 183-188.

74. **Ollivier G., Brutus L., Cot M.**, (1998). Intestinal schistosomiasis caused by *S. mansoni* in Madagascar: extent and distribution of the endemicity. *Bull. Soc. Path. Exo.*, **92** (2): 99-103.

75. **WHO**, (1993). WHO Manual. Analysis of Human Semen and the Interaction of Spermatozoa with Cervical Mucus. Ed. INSERM, Paris, 125p.

76. **WHO**, (1994). Chart for the diagnosis of intestinal parasites. Chart 3.

77. **Pagès J.R., Jourdane J., Southgate V.R., Tchuem Tchuenté L.A.**, (2003). Recognition of two sister species within the taxon *Schistosoma intercalatum* Fisher, 1934, the causative agent of human rectal schistosomiasis in Africa. Description of *Schistosoma guineensis* N. sp. Taxonomy, Ecology, and Evolution of Parasitic Metazoans. (Tribute Volume to Louis Euzet). Volume II. Combes C & Jourdane J. (eds). PUP, Perpignan, France, pp. 139–146.

78. **Pamo T.E., Tapondjou L., Tendonkeng F., Nzogang J.F., Djoukeng J., Ngandeu F., Kana J.R.**, (2003). Effect of essential oils from the leaves and flowering tips of *Cupressus lusitanica* on the tick (*Rhipicephalus lunulatus*) in Western Cameroon. *Journal of the Cameroon Academy of Sciences*, **3** (3): 169–175.

79. **Patil P.S., Elem B.**, (1988). Schistosomiasis of the prostate and the seminal vesicles: Observations in Zambia. *Journal of Tropical Medicine and Hygiene*, **91** (5): 245-248.

80. **Prendergast F.G., Veneziale C.M.**, (1975). Control of fructose and citrate synthesis in guinea pig seminal vesicle epithelium. *Journal of Biological* Chemistry, **25**: 1282–1289.

81. **Ratard R.C., Kouemeni L.E., Ekani Bessala M.M., Damkou C.N., Greer G.J., Spilsbury J., Cline B.L.** (1990). Human schistosomiasis in Cameroon. Distribution of schistosomiasis. The *American Journal of Tropical Medicine and Hygiene,* **42** (6): 561-572.

82. **Rashida B., Hala E., Alan F.**, (2005). Efficacy of Myrrh in the treatment of human *Schistosoma mansoni*. The *American Journal of Tropical Medicine and Hygiene*, **73** (2): 365-367.

83. **Ripert C.**, (1998). Epidemiology of parasitic diseases. Protozoan and helminthic infections, reservoirs, vectors, and transmission. Volume II. Helminthic infections. Editions Médicales Internationales. ISBN: 2-7430-0091-0. pp. 1–271.

84. **Rizk A.M., Abdel Kader M.M., Hashmat H.A., El-Agouz W., Abdella M.**, (1980). Sex steroids in schistosomal liver disease. *Acta Biologica et Medica Germanica*, **39**: 991-993.

85. **Robel P.**, (1991). Steroidogenesis: Enzymes and the Regulation of Their Genomic Expression. In: Reproduction in Mammals and Humans; **Thibault M.C.** (Eds). Edition Marketing, Paris, 127–134.

86. **Saez J.M., Lejeune H., Avallet O., Harbert R., Durand P.**, (1995). Control of the differentiated functions of Leydig cells. *Medicine/Sciences*, **11**: 547-553.

87. **Saleh S., Hamdy M.A.**, (1979). Testicular lipids and serum FSH in mice with experimental *Schistosomiasis mansoni. Egypt.*

88. *J. Bilharz.*, **6** (1-2): 51-9.

89. **Samuels L.T., Harding B.W.**, (1962). Aldose Reductase and Ketose Reductase in Male Accessory Reproductive Organs. *Biochemistry* Journal, 39-84.

90. **Sellin B., Boudin C.**, (1981). Schistosomiasis in West Africa. *Medical Studies*, **1**: 3-86.

91. **Setchell B.P., Brooks D.E.**, (1988). Anatomy, vasculature, innervation, and fluid of the male reproductive tract. In: The physiology of reproduction. **Knobil E., Neil J.D.** (Eds.). Raven Press, N.Y., pp. 753–836.

92. **Sheweita S.A., Mangoura S.A., El-Shemi A.G.**, (1998). Different levels of *schistosomiasis and Schistosoma mansoni* infection-induced changes in drug-metabolizing

enzymes. *Journal of Helminthology,* **72**: 71–77.

93. **Silva L.M., Menezes R.M., De Oliveira S.A., Andrade Z.A.**, (2003). Chemotherapeutic effects on larval stages of *Schistosoma mansoni* during infection and reinfection of mice. *Rev Soc Bras Med Trop,* **36**: 335-341.

94. **Siegrist D., Siegrist-Obimpeh P.**, (1992). *Schistosoma haematobium* infection in pregnancy. *Acta Tropica*, **50**: 317-321.

95. **Skelly P.J., Secor W.E., Reis M.G., Ramos E.A., Carno T.M., Peixoto E.E., Harn D.A.**, (1994). Failure of schistosomiasis to significantly decrease testosterone levels in Brazilian men. *American Journal of Tropical Medicine and Hygiene*, **51** (1): 40-44.

96. **Skelly P.J., Tielens A.G.M., Shoemaker C.B.**, (1998). Glucose transport and metabolism in mammalian-stage schistosomes. *Parasitology Today*, **14** (10): 402-406.

97. **Southgate V.R., Jourdane J., Tchuente L.A.T.**, (1998). Recent studies on the reproductive biology of schistosomes and their relevance to speciation in the Digenea. *International Journal of Parasitology*, **28** (8): 1159-1172.

98. **Stahler M.S., Colin B.G., Pansky B.**, (1987). Evidence for insulin synthesis in normal mouse seminal vesicles based on *in situ* RNA-DNA hybridization. *Biology and Reproduction* **36**: 999-1006.

99. **Steinberger R.M., Lindsay K.G., Alassandri R., Wise G.J.**, (1975). Testicular infarction and *Schistosoma mansoni. Urology*, **5** (4): 467-496.

100. **Tchuem Tchuenté L.A.**, (1993). Host-parasite interactions and interspecific competition in the *Schistosoma intercalatum* model: Epidemiological implications. Doctoral thesis, University of Perpignan. 165p.

101. **Tendler M., Pinto R.M., Oliviera L.A., Gebara G., Katz N.**, (1986). *Schistosoma mansoni* vaccination with adult worm antigens. *International Journal of Parasitology*, **16**: 347-352.

102. **Théron A., Mone H., Gerard C.**, (1992). Special and energy trade-offs between host and parasite: the *Biomphalaria glabrata-Schistosoma mansoni* system. *International Journal of Parasitology,* **22**: 91–94.

103. **Tiboldi T.**, (1979). Ovaries and Adrenal Glands in Murine Schistosomiasis mansoni. I.

Histopathological Changes in the Ovaries During Acute and Chronic Infection. *American Journal of Tropical Medicine and Hygiene*, **28** (4): 670–676.

104. **Trease G.E., Evans W.C.**, (1983). Pharmacognosy. Ballière Tindall Press. London, pp. 309–706.

105. **Venkatesh S., Reddy Y.S., Suresh B., Reddy B.M., Ramesh M.**, (1999). Antinociceptive and anti-inflammatory activity of *Sida rhomboidea* leaves. *J. Ethnopharmacol.*, **67** (2): 229–32.

106. **Vilana R., Corachan M., Gascon J., Valls E., Bru C.**, (1997). Schistosomiasis of the female genital tract: transrectal sonographic findings. *Journal of Urology*, **4**: 1491–1493.

107. **Wasilewski M.M., Lim K.C., Phillips J., McKerrow J.H.**, (1996). Cysteine protease inhibitors block schistosome hemoglobin degradation *in vitro* and decrease worm burden and egg production *in vivo*. *Molecular and Biochemical Parasitology,* **81** (2): 179-89.

108. **WHO**, (1997). CTD Programme Summary. Division of Control of Tropical Diseases, 9p.

109. **Woolhouse M.E.J., Chandiwana S.K.**, (1990). Population biology of the freshwater snail *B. globosus* in the Zimbabwean highveld. *Journal of Applied Ecology*, **27**: 41-59.

APPENDICES

Appendix 1: Average load of *Schistosoma mansoni* eggs per gram of feces before and after treatment of infested mice

Groups	Number	Extract dose (mg/kg)	Mean number of eggs per gram of feces ± SEM		P (%)	Impact of treatment
			Before treatment (^{60th}day or^{8th}week)	After treatment (81stweekor^{11th}week)		
Tn	10	0	0	0	0	
Tp60	10	0	595.2	/	/	
Tp81	10	0	242.40 ±	271.20 ±	+11.88	**TP**: 271.2
E40	10	40	61.70	56.80	+96.43	± 56.80
E80	11	80	134.40 ±	264.00 ±	-58.33	
E160	12	160	34.35	57.80	-43.60	**AT**:
E320	11	320	130.91 ±	54.55 ±	+16.95	123.33
			48.04	24.02		± 54.07
			43.00 ±	24.25 ±		
			6.52	24.25***		
			128.73 ±	150.55 ±		
			34.29	48.55		

P: percentage increase or decrease in the number of eggs per gram of feces in treated mice compared to control mice

*** $p < 0.001$ significant compared to the number of eggs before treatment

SEM: standard error of the mean

Tn: negative controls, Tp60: untreated parasitized animals sacrificed on the^{60th}day; Tp81: untreated parasitized animals sacrificed on the 81st^{e} day, E40: animals treated at a dose of 40 mg/kg, E80: animals treated at a dose of 80 mg/kg, E160: animals treated at a dose of 160 mg/kg, E320: animals treated at a dose of 320 mg/kg, TP: parasitized animals, AT: treated animals.

Appendix 2: Average load of *Schistosoma mansoni* eggs per gram of liver

Groups	Number of animals	Extract dose (mg/kg)	Mean number of eggs per gram of liver ± SEM	Impact of treatment
Tn	10	0	0	
Tp60	10	0	5,236.32 ± 1,142.95	
Tp81	10	0	5432.39 ± 702.12	**TP**: 5334.35 ± 98.04
E40	10	40	8063.49 ± 1486.08	
E80	11	80	3408.82 ±	
E160	12	160	1209.23$^{\$}$	**AT**: 4121.30 ±
E320	11	320	164.42 ± 164.42$^{\$\$\$}$	1638.96
			4848.48 ± 532.20	

$ ($p < 0.05$), $$$ ($p < 0.001$) significant difference compared to untreated, parasitized animals sacrificed on the^{81st}day

SEM: standard error of the mean

Tn: negative controls, Tp60: untreated parasitized animals sacrificed on the^{60th}day, Tp81: untreated parasitized animals sacrificed on the 81stday, E40: animals treated with a dose of 40 mg/kg, E80: animals treated with a dose of 80 mg/kg, E160: animals treated with a dose of 160 mg/kg, E320: animals treated with a dose of 320 mg/kg, TP: parasitized animals, AT: treated animals.

Appendix 3: Mean number of schistosomes in infested mice and mice treated with aqueous extract of *Clerodendrum umbellatum* leaves

Groups	Number	Extract dose (mg/kg)	Average number of schistosomes ± ESM	Impact of treatment
Tn	5	0	0	
Tp60	5	0	40 ± 13.12	
Tp81	5	0	32.60 ± 22.79	**TP**: 36.3 ± 3.70
E40	5	40	8.60 ± 4.89	
E80	6	80	3.67 ± 2.54	
E160	6	160	0.00 ± 0.00[$$]	**AT**: 4.19 ± 1.76[&&&]
E320	6	320	4.50 ± 3.39	

$$ p<0.05 significant difference compared to untreated parasitized animals sacrificed on day[81](Tp81 group)

SEM: standard error of the mean

Tn: negative controls, Tp60: untreated parasitized animals sacrificed on the[60th]day, Tp81: untreated parasitized animals sacrificed on the 81st[e] day, E40: animals treated with a dose of 40 mg/kg, E80: animals treated with a dose of 80 mg/kg, E160: animals treated with a dose of 160 mg/kg, E320: animals treated with a dose of 320 mg/kg, TP: parasitized animals, AT: treated animals.

Appendix 4: Changes in average body weights of mice during the 11-week experiment

Groups							
Tn	**Tp60**	**Tp81**	**E40**	**E80**	**E160**	**E320**	**Weeks**
100 ± 0.00	100 ± 0.00	100 ± 0.00	100 ± 0.00	100 ± 0.00	100 ± 0.00	100 ± 0.00	**1**
100.89 ± 0.62	97.9 ± 2.13	112.23 ± 2.96	108.05 ± 1.73	102.6 ± 1.82	104.97 ± 1.21	111.78 ± 1.85	**2**
103.31 ± 0.81	112.41 ± 3.03	117.41 ± 3.30	110.4 ± 4.16	104.22 ± 3.78	107.96 ± 2.16	120.14 ± 2.28	**3**
108.87 ± 1.63	117.56 ± 2.87	120.89 ± 2.86	110.37 ± 3.38	105.29 ± 3.49	111.43 ± 1.78	125.20 ± 3.20	**4**
96.45 ± 1.19	118.73 ± 3.46	126.31 ± 2.83	112.6 ± 3.85	107.3 ± 3.58	113.45 ± 5.13	123.17 ± 2.66	**5**
96.68 ± 1.07	125.49 ± 3.52	133.07 ± 4.49	114.63 ± 3.47	108.95 ± 3.82	114.61 ± 4.35	125.45 ± 2.55	**6**
102.07 ± 1.45	136.85 ± 5.84	125.36 ± 4.69	117.79 ± 3.56	106.63 ± 3.60	114.2 ± 5.61	125.73 ± 3.19	
101.37 ± 1.22	135.15 ± 5.94	127.39 ± 4.34	119.28 ± 3.33	112.32 ± 4.40	113.23 ± 4.66	125.63 ± 3.93	**7**
105.11 ± 1.05	117.16 ± 3.55	121.19 ± 5.77	120.5 ± 4.52	104.86 ± 4.75	96.8 ± 3.62	127.63 ± 3.47	**8**
104.05 ± 1.01		117.48 ± 4.29	115.64 ± 4.23	100.95 ± 5.05	102.23 ± 3.40	118.51 ± 3.33	**9**
105.22 ± 1.03		117.96 ± 4.64	112.4 ± 4.54	97.64 ± 5.42	100.39 ± 3.03	125.99 ± 3.36	**10**
108.62 ± 1.15		118.78 ± 4.26	110.3 ± 4.89	96.33 ± 5.55	102.21 ± 3.69	101.08 ± 2.77	**11**

Values represent the mean ± SEM (standard error of the mean)

Tn: negative control animals, Tp60: untreated parasitized animals sacrificed on the[60th]day, Tp81: untreated parasitized animals sacrificed on the 81st[e] day, E40: animals treated with a dose of 40 mg/kg, E80: animals treated with a dose of 80 mg/kg, E160: animals treated with a dose of 160 mg/kg, and E320: animals treated with a dose of 320 mg/kg

<u>Appendix 5</u>: Variation in mean relative weights of the liver and spleen

Groups	Extract dose (mg/kg)	Mean relative weights of the liver ± SEM	Mean relative spleen weights ± SEM	Impact of treatment	
				Liver	Spleen
Tn	0	4.76 ± 0.10 (10)	0.5 ± 0.08 (5)		
Tp60	/	9.01 ± 0.56*** (10)	0.84 ± 0.10* (5)	**TP**:	**TP**:
Tp81	0	9.13 ± 0.29*** (10)	1.03 ± 0.05*** £ (5)	9.06 ±	0.93 ±
E40	40	6.55 ± 0.33$$$ (10)	0.63 ± 0.08$ (5)	0.06	0.09
E80	80	5.87 ± 0.56$$$ (11)	0.36 ± 0.03$$$ (5)		
E160	160	4.99 ± 0.31$$$ (12)	0.43 ± 0.12$$$ (6)	**AT**:	**AT**:
E320	320	7.29 ± 0.37$$ (11)	1.35 ± 0.35(5)	6.17 ±	0.69 ±
				0.49&	0.29

* p<0.05; ***p<0.001 significant difference compared to negative control animals (Tn) $ p<0.05; $$ p<0.01; $$$ p<0.001 significant difference compared to untreated parasitized animals (Tp81)

SEM: standard error of the mean

Tn: negative control animals, Tp60: untreated parasitized animals sacrificed on the^{60th}day, Tp81: untreated parasitized animals sacrificed on the 81st^{e} day, E40: animals treated with a dose of 40 mg/kg, E80: animals treated with a dose of 80 mg/kg, E160: animals treated with a dose of 160 mg/kg, E320: animals treated with a dose of 320 mg/kg, TP: parasitized animals, AT: treated animals.

Appendix 6: Variation in mean relative weights of the epididymis, seminal vesicles, and testes

Groups	Number of animals	Extract dose (mg/kg)	Mean relative weights of epididymis ± SEM	Mean relative weights of seminal vesicles ± SEM	Mean relative weights of testes ± SEM
Tn	5	0	0.12 ± 0.01	0.53 ± 0.05	0.498 ± 0.03
Tp60	5	/	0.18 ± 0.02*	0.62 ± 0.13	0.54 ± 0.03
Tp81	5	0	0.07 ± 0.01*** £	0.44 ± 0.11	0.332 ± 0.07£
E40	5	40	0.17 ± 0.01$	0.52 ± 0.08	0.598 ± 0.06$
E80	5	80	0.20 ± 0.01 $$$	0.79 ± 0.13$	0.552 ± 0.04$
E160	6	160	0.13 ± 0.01$$$	0.61 ± 0.08	0.54 ± 0.05$
E320	5	320	0.21 ± 0.02$$$	0.41 ± 0.07	0.61 ± 0.05$$

* $p<0.05$; ***$p<0.001$ significant difference compared to negative control animals

£ $p<0.05$ significant difference compared to untreated parasitized animals sacrificed on the 60th day

$ $p<0.05$; $$ $p<0.01$; $$$ $p<0.001$ significant difference compared to untreated parasitized animals sacrificed on the 81st day

SEM: standard error of the mean

Tn: negative control animals, Tp60: untreated parasitized animals sacrificed on the 60thday, Tp81: untreated parasitized animals sacrificed on the 81st day, E40: animals treated with a dose of 40 mg/kg, E80: animals treated with a dose of 80 mg/kg, E160: animals treated with a dose of 160 mg/kg, and

E320: animals treated with a dose of 320 mg/kg

Appendix 7: Variation in epididymal alpha-glucosidase activity (mU/g of organ)

Groups	Number of animals	Extract dose (mg/kg)	Epididymal alpha-glucosidase activity (mU/g of organ)
Tn	5	0	0.512 ± 0.03
Tp60	5	0	0.428 ± 0.04
Tp81	5	0	0.41 ± 0.02*
E40	5	40	0.624 ± 0.02$$$
E80	5	80	0.508 ± 0.04$
E160	6	160	0.588 ± 0.02$$$
E320	5	320	0.378 ± 0.02

Values represent means ± SEM (standard error of the mean)

* p<0.05: significant difference compared to negative control animals

$ p<0.05; $$$ p<0.001 significant difference compared to untreated parasitized animals sacrificed on the81stday

Tn: negative control animals, Tp60: untreated infected animals sacrificed on day60, Tp81: untreated infected animals sacrificed on day 81day, E40: animals treated with a dose of 40 mg/kg, E80: animals treated with a dose of 80 mg/kg, E160: animals treated with a dose of 160 mg/kg, and E320: animals treated with a dose of 320 mg/kg.

Appendix 8: Average concentration of seminal fructose

Groups	Number of animals	Extract dose (mg/kg)	Seminal fructose concentration (µmol/g of organ)
Tn	5	0	2.046 ± 0.40
Tp60	5	0	2.836 ± 0.21*
Tp81	5	0	2.54 ± 0.06
E40	5	40	2.78 ± 0.41
E80	5	80	1.79 ± 0.13$$
E160	6	160	1.92 ± 0.19$
E320	5	320	1.71 ± 0.07$$

Values represent means ± SEM (standard error of the mean)

* $p<0.05$: significant difference compared to negative control animals

\$ $p<0.05$; \$\$ $p<0.01$ significant difference compared to untreated parasitized animals sacrificed on the 81st day

Tn: negative control animals, Tp60: untreated parasitized animals sacrificed on the 60th day, Tp81: untreated parasitized animals sacrificed on the 81st[e] day, E40: animals treated with a dose of 40 mg/kg, E80: animals treated with a dose of 80 mg/kg, E160: animals treated with a dose of 160 mg/kg, and E320: animals treated with a dose of 320 mg/kg.

<u>Appendix 9</u>: Mean sperm count in control mice and mice treated with aqueous extract of *Clerodendrum umbellatum* leaves

Groups	Number of animals	Extract dose (mg/kg)	Mean sperm count ± SEM
Tn	5	0	375.00 ± 53.38
Tp60	5	0	564.40 ± 27.29*
Tp81	5	0	206.40 ± 78.63££
E40	5	40	659.60 ± 106.87$$
E80	5	80	374.20 ± 55.55
E160	6	160	390.83 ± 82.28
E320	5	320	179.20 ± 22.73

* $p<0.05$ significant difference compared to negative control animals

££ $p<0.01$ significant difference compared to untreated infected animals sacrificed on the 60th day $$ $p<0.05$ significant difference compared to untreated infected animals sacrificed on the 81st day SEM: standard error of the mean

Tn: negative control animals, Tp60: untreated parasitized animals sacrificed on the 60th day, Tp81: untreated parasitized animals sacrificed on the 81stday, E40: animals treated with a dose of 40 mg/kg, E80: animals treated with a dose of 80 mg/kg, E160: animals treated with a dose of 160 mg/kg, and E320: animals treated with a dose of 320 mg/kg.

More Books!
yes
I want morebooks!
Buy your books fast and straightforward online - at one of world's fastest growing online book stores! Environmentally sound due to Print-on-Demand technologies.
Buy your books online at
www.morebooks.shop
Kaufen Sie Ihre Bücher schnell und unkompliziert online – auf einer der am schnellsten wachsenden Buchhandelsplattformen weltweit! Dank Print-On-Demand umwelt- und ressourcenschonend produzi ert.
Bücher schneller online kaufen
www.morebooks.shop
info@omniscriptum.com
www.omniscriptum.com
OMNIScriptum

MIX
Papier aus verantwortungsvollen Quellen
Paper from responsible sources
FSC® C105338

Printed by Books on Demand GmbH, Norderstedt / Germany